THE TELESCOPE

THE TELESCOPE
A Short History

Richard Dunn

© National Maritime Museum, 2009

First published in 2009 by the National Maritime Museum, Greenwich, London SE10 9NF

This edition published in Great Britain in 2011 by Conway,
an imprint of Anova Books Ltd
10 Southcombe Street
London
W14 0RA
www.conwaypublishing.com
www.anovabooks.com
www.nmm.ac.uk

Distributed in the U.S. and Canada by:
Sterling Publishing Co., Inc.
387 Park Avenue South
New York, NY 10016-8810

All rights reserved. No part of this publication may be reproduced, stored in a data retrieval system, or transmitted in any form or by any means electronic, mechanical, photocopying, recording or otherwise without the prior written permission of the copyright owner.

A CIP catalogue record of this book is available from the British Library.

ISBN 9781844861477

Printed and bound by 1010 Printing Ltd, China

CONTENTS

Introduction	7
1. Before the telescope	11
2. The birth of the telescope 1608–1700	21
3. The telescope and the imagination 1608–1700	44
4. The reflecting telescope 1610–1800	55
5. Perfecting the refractor 1700–1800	72
6. Global domination 1720–1900	83
7. The telescopic view 1720–1900	109
8. Telescopes, binoculars and modern life	127
9. Modern astronomical telescopes	140
Epilogue – Looking to the future	164
The telescope – a short timeline	168
Glossary	170
Further reading	177
Acknowledgements	181
Picture credits	182
Index	184

1 Frontispiece from Giovanni Battista Riccioli, *Almagestum novvum* (Bologna, 1660). Riccioli, covered in eyes, holds a telescope that reveals the truth. Urania, muse of astronomy, balances two world systems, with Riccioli's the victor. Keeping the Earth at the universe's centre, his model suited Catholic orthodoxy and remained official doctrine for many years.

INTRODUCTION

Telescope, *n*. A device having a relation to the eye similar to that of the telephone to the ear, enabling distant objects to plague us with a multitude of needless details.

Ambrose Bierce, *The Devil's Dictionary*

A telescope is a simple thing. A dictionary describes it as no more than 'an optical instrument for making distant objects appear nearer and larger'. But this apparent simplicity hides an inspiring story of what has become an iconic piece of technology. The telescope was the first instrument to extend one of the human senses. It is no surprise that its impact was profound.

This book tells the story of the telescope's transformation from a simple idea into a host of instruments large and small, and of the men and women who guided these changes. It is about the ways in which people adopted, adapted and responded to such a versatile device, and its social and cultural impact.

When first turned to the skies, the telescope led to discoveries that changed the way people thought about the universe and humanity's place within it. Since that time astronomical telescopes have become impossibly large and are now able to 'see' in ways its first users could never have imagined. Modern telescopes do not just enhance the view of nearby objects, but allow astronomers to look into apparent emptiness, producing extraordinary images of the far reaches of space and the beginnings of time (see Fig. 58).

The telescope's effects have been astounding, but it has not always been a smooth ride. As an instrument that revealed things previously invisible to the eye, the telescope raised questions about the images it produced. For some, the first observations showed that the Earth and planets orbited around the Sun; for others it kept humanity at the centre of creation (Fig. 1). Debates about how to interpret the revelations of the telescope have continued for four centuries, at times treated somewhat light-heartedly and satirically (Fig. 2).

---

2 Samuel Ireland, after William Hogarth, *Royalty, Episcopacy and Law*, late eighteenth century (from an original print of *c.* 1724/5). Inspired by a recent lunar eclipse, Hogarth shows the Moon's 'inhabitants' revealed by a telescope. The truth his telescope uncovers is England's corruption: a coin for the king's face, and a pump pouring cash into a bishop's money-chest.

Some of the Principal Inhabitants of the MOON, as they were Perfectly Discover'd by a Telescope brought to y^e Greatest Perfection since y^e last Eclipse Exactly Engraved from the Objects, whereby y^e Curious may Guess at their Religion Manners. &c.

3 'Glass-grinder's roundelay' from William Kitchiner, *The Economy of the Eyes Part II* (London, 1825). Kitchiner was interested in lens-making because good telescopes needed good lenses. He hoped that singing would improve production rates.

Alongside this is a story on a smaller scale, about mass production. This is the tale of an instrument used everywhere to get a better view – on ships, in war, in recreation. Binoculars are a part of this story, since they have been around just as long, although they only rose to prominence in the twentieth century.

What unfolds is a tale of innovation in which money and the military crop up again and again, whether in the first attempts by Galileo and others to exploit the new invention in the early seventeenth century, or in the creation of giant telescopes with obvious surveillance capabilities in the last sixty years. There are many heroes, some more deserving than others, and many more subtle presences: it would be wrong to forget the glass-makers, lens-grinders and other workers who played a crucial role in the telescope's development, but are rarely given centre stage (Fig. 3).

The telescope's tale has many sides. It is not just about technical developments and their scientific application, but also about how people view themselves, others and the universe that surrounds them. It is a tale that deserves to be told in full.

Chapter One

BEFORE THE TELESCOPE

For we can so shape transparent bodies, and arrange them in such a way with respect to our sight and objects of vision, that . . . we shall see the object near or at a distance . . . Thus a small army might appear very large . . . So also might we cause the sun, moon, and stars in appearance to descend here below.

Roger Bacon, *Opus Maius* (about 1267)

What was the world like before the invention of the telescope? How did people use lenses and mirrors and imagine how they worked? What did they think was out there when they looked into the sky? If we try to answer these questions, we can begin to understand where the idea of making a telescope came from and why it had such an impact.

That transparent substances like rock crystal produced interesting optical effects was known in many ancient cultures, as were the reflective properties of metals and other materials. Mirror-making had a particularly long history, with examples made from polished stone dating from as much as 8,000 years ago. Lenses came along later. Craft workers in ancient Egypt used polished crystals on statues and burial caskets to add to their visual impact. Intriguing lens-like artefacts have also been unearthed in the remains of Troy (2600–2200 BC), in Crete (1600–1200 BC) and in the ruins of Nimrud in Iraq (probably from the seventh century BC), but their use is uncertain. Some centuries later, Aristophanes' play *The Clouds* (423 BC) describes burning glasses, but no record has yet been discovered of lenses being used to magnify images.

Before long ancient philosophers began to think about the behaviour of light. In the fourth century BC Aristotle questioned how it could pass through a solid crystal. He also observed that, 'the man who shades his eye with his hand or looks through a tube will...see further', so the use of long tubes as seeing aids must have been well known (Fig. 4). Following on from Aristotle's work, a coherent system of optical principles appeared in Euclid's *Optics*. Later authors expanded on Euclid's text to establish rules describing the behaviour of light, including the principles that it travels in straight lines and obeys laws of reflection. Meanwhile, mirrors began to feature in tales of powerful optical devices. According to one, the Greek mathematician Archimedes repelled an attack on Syracuse in 212 BC by setting the approaching ships alight using the Sun's rays reflected from large mirrors.

4   An astronomer using an astrolabe and a sighting tube, from a print by W.B. Rye, 1854, after a thirteenth-century French manuscript.

While glass-making was known in Egypt as early as the fourth millennium BC, it was the ancient Romans who developed techniques for making transparent, colourless glass and who first made glass lenses. Like the ancient Greeks, the Romans knew about burning glasses: in the first century AD the natural historian Pliny the Elder noted that medical practitioners cauterized wounds with 'a ball of crystal acted upon by the rays of the sun'. The Romans also knew about magnifying lenses, with the writer Seneca remarking that letters can be enlarged by looking through a glass globe filled with water.

Optical theory played a crucial role in medieval Islamic philosophy, which in turn began to re-shape European thought by the thirteenth century. Two important English thinkers, Robert Grosseteste and Roger Bacon, were part of this revival of European philosophical speculation. Grosseteste described optical devices to 'make small things placed at a distance appear any size we want'.

Before the Telescope   13

Bacon's encyclopedic *Opus Maius* (about 1267) also encompassed optical theory and described a glass sphere to aid vision as well as other wonderful instruments. But these intriguing references were probably speculative rather than specific descriptions. Other medieval works retold fantastic tales of ancient wonders. It was said that Julius Caesar used 'great Glasses' to look from France to the coast of England, and that the Pharos of Alexandria once contained an optical device that could see approaching ships from far away and set fire to them.

On a more practical level, optical lenses began to be made commercially in Europe at the end of the thirteenth century, by which time glass of reasonable quality was being produced in Venice and Florence. Eyeglasses to help people read were available in Florence by *c.*1280. Their convex lenses suited people who were long-sighted. Concave lenses were harder to produce and only became widely available in the fifteenth century. By then the advent of the printing press had dramatically increased the number of books available. As a result the number of short-sighted people soared, all needing spectacles with concave lenses. So by about 1450, convex and concave lenses were readily available, albeit of poor quality by the standards of later centuries. In the 1500s, Venice also became the centre for making mirrors from glass coated with a metal amalgam.

As lenses and mirrors became more common, their quality improved and philosophers of the sixteenth century began to speculate about the possibilities they offered. A number made bold claims about devices they were able to make, including instruments for seeing things far away. In about 1570, the Englishman Thomas Digges described one made years before by his father Leonard. He was said to be able to read letters or the inscriptions on coins from a distance, and even see what was happening 'in private places' from seven miles away.

While there is still debate as to what Digges and other writers were describing, their writings do seem to be based on real experiments with lenses and mirrors, although definitive descriptions of the instruments they made have not yet been discovered. In a similar vein, the Elizabethan court adviser John Dee wrote that the military tactician 'may wonderfully helpe him selfe by perspective Glasses', while admitting that 'posterity will prove more skillful and expert' in the matter. The possibility was there, he suggested, if not the reality. But what is important is that by the late sixteenth century many people were experimenting with lenses and mirrors and thinking about their potential.

The history of astronomy is equally complex. Observers in prehistoric times must certainly have linked the changing sky to the seasonal cycles of the agricultural year. Over the generations this basic knowledge evolved into three practical reasons for carrying out systematic observation: religious and mystical beliefs, reflected in the alignment of monuments and tombs; measuring time, including the identification of prayer times and significant days; and prognostication or prediction, particularly astrology.

By late antiquity, ancient Greek philosophers were trying to understand the nature of the universe and explain the motions of the stars and planets. It was Aristotle who pulled together the work of his predecessors and established a set of rules that were still the dominant view in the early seventeenth century. According to his world-view, the Earth lay at the centre of a finite universe, with the heavenly bodies moving in circles around it. He also maintained a fundamental distinction between the earthly and heavenly regions. The terrestrial sphere was made of four elements (earth, water, air and fire), which moved in straight lines and were subject to change and decay. The stars and planets were made of a quite different fifth element. This was perfect and unchanging, and moved only in circles. As these principles were further elaborated over the next two millennia, the idea took hold that the spheres in which the planets moved were substantial, perhaps of a crystalline nature.

At the same time, Greek astronomers were working to produce mathematical models that would give accurate predictions of the future positions of the stars and planets. This culminated in the second-century writings of Ptolemy, in particular the *Mathematike syntaxis* or 'Mathematical Compilation', generally known as the *Almagest* from the Arabic for 'the greatest'. Ptolemy's magisterial work brought together and systematized ancient Greek thinking about the cosmos. But it also revealed a dilemma. On the one hand, Aristotle's philosophical principles of uniform circular motion around a stationary Earth were simple. On the other, the mathematical models required to produce accurate predictions of the planets' actual motions were much messier.

Ptolemy's mathematical astronomy was still based on motions in circles, but not in the simple way Aristotle imagined. It included complex compound motions of circles on circles, some no longer centred on the Earth, others with a non-uniform motion. His descriptions were accurate for the purposes of prediction but were not philosophically pure. There was no obvious way of resolving this apparent tension between a philosophical ideal and computational necessity, so the two co-existed in an uneasy alliance.

Islamic scholars made significant advances in both philosophical and mathematical astronomy in the middle ages. By this time the need to update Ptolemy's astronomical data led to the foundation of dedicated observatories to produce new star catalogues. Islamic astronomers and artisans also made improvements to instruments such as the astronomical quadrant and astrolabe (Fig. 5). In the same period Islamic philosophers started to question the physical reality of the Aristotelian–Ptolemaic world system. As their works and translations of ancient Greek texts were transmitted back to Europe from about the twelfth century, similar criticisms emerged among European philosophers, but the predictive power of Ptolemy's mathematical models meant that astronomers still needed them.

The same problems and questions were still troubling scholars in the sixteenth century. Their worries were compounded as the calendar was known to be increasingly out of step with the seasons

5   Astronomical astrolabe, by Maḥmūd ibn Shawka al-Baghdādī, *c.* 1294–95.

and the dating of Easter (a problem that lay behind the creation of the Gregorian calendar). Then in the early years of the century a canon of the cathedral of Frombork (Frauenberg) named Nicolaus Copernicus set out his thoughts on the matter. A manuscript circulated to other scholars argued that a Sun-centred model of the universe would be simpler and more accurate than the Ptolemaic approach. After some encouragement, he was persuaded to elaborate his ideas, which were finally published in *De revolutionibus orbium coelestium* (1543). Copernicus proposed that the Sun was the centre of the universe, and that the Earth, planets and stars revolved around it in circular orbits (Fig. 6). He also worked through the mathematics required to model the planetary motions in the new system. More radically, he tackled the question of the relationship between the philosophical ideal and mathematical accuracy. A true

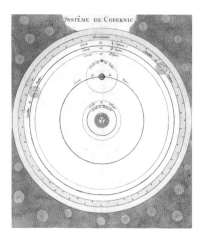

6 The Copernican world-system, from a French engraving, 1761.

description of the universe, he said, must be both philosophically tenable and mathematically accurate. He believed that his heliocentric model did just that.

Copernicus's proposal was of great interest, but most astronomers simply considered it as a mathematical calculating device for making accurate predictions. They saw no need to consider its philosophical claims until thirty years later, when two astronomical events offered significant challenges to the Aristotelian description of the universe. The first was the appearance in 1572 of a new star in the constellation of Cassiopeia, which flew in the face of Aristotle's dictum that the heavenly regions could not change. Five years later astronomers tracked a comet seen from Europe and determined that it too was in the heavenly regions. Again, this contradicted the accepted view that these regions were made of crystalline spheres. How could a comet pass through solid matter, astronomers asked?

Among those who observed and wrote about the phenomena of 1572 and 1577 was the larger than life figure of Tycho Brahe. Born into a Danish noble family, Tycho spent the years until he was about thirty at various European universities. During this time he developed a passion and talent for astronomy, spending a fortune on books and instruments. It was also in this period that he lost part of his nose in a duel. This left him with a rather distinctive appearance with a metallic nose-piece to which he was often seen applying ointment.

'I began to doubt the faith of my own eyes', Tycho later recalled of his first glimpse of the new star of 1572. Further observations convinced him that it was proof that Aristotle's world-view was seriously flawed. If Aristotle was wrong, he asked, what was the true arrangement of the universe? An obvious candidate was Copernicus's model, but how could he know for sure? Perhaps the best way was to obtain accurate observational data.

Tycho therefore oversaw the foundation of Europe's first dedicated observatory on the Danish island of Hven. With the first buildings completed in 1580, his subsequent work there revolutionized the practice of astronomy. Above all, it established the importance of accurate observation with high-quality instruments. He used naked-eye observation through pinhole sights, with angular measurements made against large, accurately engraved scales. It was to be another century before astronomers achieved greater accuracies.

Meanwhile, the results began to change people's thinking about the universe and its operations. Tycho himself soon realised that the current world-system was most definitely incorrect: by 1588 he was unequivocal. 'There are no solid spheres in the heavens', he wrote. Thinking to test the Copernican model, he tried to detect stellar parallax, the apparent change in the relative positions of objects due to the observer's changing position. Parallax should automatically be observed if, as Copernicus claimed, the Earth moved around the Sun. Finding none, Tycho rejected the moving-Earth model and put forward an alternative in which the Sun and Moon revolved around a stationary Earth, while the remaining planets went around the Sun. This not only fitted Tycho's observations, including the lack of parallax, but also maintained Earth's prestigious place at the centre of creation (see Fig. 1). For many, it was the best alternative even into the following century, when crucial evidence for the Earth's motion around the Sun was finally observed. Even so, Tycho's system still managed to be defended as late as 1857, albeit in a rather eccentric book, *The Solar System As It Is, and Not As It Is Represented*, by Lieutenant R.J. Morrison, RN, also known as the astrologer Zadkiel.

Tycho's observational results proved of even greater value in the hands of Johannes Kepler, who became his assistant in 1600. By this time Tycho was Imperial Mathematician at the Prague court of the Holy Roman Emperor Rudolf II, and knew of Kepler's talent as a mathematician. Unfortunately, the two never got on, so it was only on his deathbed that Tycho handed his results to Kepler, acknowledging that he was the best person to make use of them. Kepler took over as Imperial Mathematician and began to wrestle with a mathematical model to explain Tycho's observations of Mars's orbit, which was known to deviate significantly from the 'ideal' of perfect circular motion. The work took him many years but paid rich dividends in a new astronomical system in which the planets moved in elliptical orbits around a central Sun. Kepler would also play an important role in the early history of the telescope; somewhat ironically, since he had very poor eyesight as a result of a childhood bout of smallpox.

As the telescope's story began, much was in flux. The long-accepted geocentric view of the universe was subject to increasing criticism, both philosophically and from recent astronomical observations. Yet it was by no means deposed and remained central to Christian doctrine and university teaching. A new impetus for accuracy in observational astronomy was also growing, promoting the use of accurate instruments to reveal the celestial movements, with a view to finally resolving questions about the physical nature of the universe. At the same time, scholars and artisans were experimenting with lenses and mirrors and were becoming convinced of the exciting possibilities they offered. It was against this background that the story of the telescope was to play out.

Chapter Two

THE BIRTH OF THE TELESCOPE 1608–1700

O telescope, instrument of much knowledge, more precious than any sceptre! Is not he who holds thee in his hand made king and lord of the works of God?

Johannes Kepler, *Dioptrice* (1611)

Galileo was not the first: neither the first to build a telescope, nor the first to turn it to the heavens. Yet his name became synonymous with the invention and it is Galileo who is remembered today because of what he achieved with it. He deserves fame, but not exclusively so; many others were involved in the creation and development of an instrument that became a common optical aid and a tool of discovery.

The true inventor of the telescope may never be known. Surviving records lead to The Hague in September 1608. This was a difficult time for the peoples of the United Provinces of the Dutch Republic. They had been fighting to gain independence from Spain since 1568, and The Hague was now the site of a conference held to negotiate an end to the hostilities; English, French and Belgian delegates attended alongside observers from other countries. Talks had been going on for eight months but had recently split up in deadlock.

In the middle of these tense negotiations, a young man arrived from Zeeland in the south to see Maurice of Nassau, commander of the Republic's army. The man was carrying a letter saying that he was the inventor of a new instrument, 'by means of which all things at a very great distance can be seen as if they were nearby'. It was a tube containing two glass lenses.

This may seem a strange occurrence at peace talks, but the military implications of the invention must have been obvious. The States-General formed a committee to examine the device and the application for a patent from Hans Lipperhey (d. 1619), a spectacle-maker working in Middelburg. But by mid-October two other men were claiming to have invented, or at least to be able to make, the new apparatus. One was Jacob Adriaenszoon (better known as Jacob Metius) of Alkmaar, the other probably Sacharias Janssen, another spectacle-maker from Middelburg. With the secret already widely known, the States-General decided against awarding a patent to any of the claimants.

But were any of these men the first to invent the telescope? By the late sixteenth century a number of authors were already describing optical instruments based on their own experiments. It has been suggested that Leonard Digges in England made a telescope with a lens and a mirror sometime around the 1550s, the instrument his son Thomas later wrote about. The evidence remains uncertain, but even if true, Digges's trials did not result in a mass-produced device. What is clear is that that many people were making investigations with lenses of sufficient quality for a working telescope to be a practical possibility.

It is safe to say that the telescope's invention predates 1608, with the key discovery perhaps made by chance as someone was carrying out investigations or even just playing around with lenses and chanced upon the right combination. Perhaps they noticed a small magnifying effect but did not see any significant potential for it. Lipperhey's announcement in 1608 was a breakthrough because he was among the first to realize the device's possibilities and to seek to exploit them. Even if he wasn't the first to discover the correct lens combination, he deserves credit for appreciating its importance.

Whoever invented the new instrument, it certainly aroused interest in The Hague, where the soldiers from Europe recognized the value of something that could help them spot their enemies from a distance. According to one early report,

> The said glasses are very useful in sieges and similar occasions, for from a mile or so away one can detect all things as distinctly as if they were very close to us. And even the stars . . . can be seen by means of this instrument.

The same report claimed that Maurice of Nassau showed the new contraption to Ambrogio Spinola, commander of the Spanish forces, who responded that, 'I could no longer be safe, for you will see me from afar'. Pierre Jeannin, head of the French negotiators, was equally impressed and asked Lipperhey to make two more to send to France. But Lipperhey had already promised not to make any more. Yet despite the States-General's initial attempts to keep it

7   Galileo Galilei (1564–1642), by Justus Sustermans, *c.* 1639. Painted a few years before Galileo's death, the portrait appropriately shows him holding a telescope.

under wraps, copies of the device spread across Europe within months. By spring 1609 spectacle-makers were selling small spyglasses on the streets of Paris. Lipperhey's military innovation was becoming commercial.

This is where Galileo Galilei comes in (Fig. 7). Originally from Pisa, Galileo and his family moved to Florence in the early 1570s while he was a young boy. At the age of seventeen he enrolled at the University of Pisa, where his father hoped he would study the useful and lucrative subject of medicine. But the young man became more interested in mathematics and mechanics. As his interest blossomed, he resolved to make a career of it and was appointed Professor of Mathematics at the University of Padua in the Venetian Republic in 1592.

Financially, this was no bad thing. Galileo was now head of the family following the death of his father. In Padua he also began a relationship with Marina Gamba, with whom he had three children.

With a growing number of dependents and a penchant for drinking and dining, before long the university professor had to supplement his income by giving private tuition and selling mathematical instruments.

According to Galileo's own account, news of a spyglass, 'made by a certain Dutchman', reached him in Padua in about May 1609. Given the speed with which reports first circulated in Europe, this seems rather late, especially since rumours certainly reached Paolo Sarpi, a Paduan acquaintance of Galileo's, by November 1608. It seems more likely that Galileo heard some brief report of the instrument at this earlier date, but assumed that it was based on a lens–mirror combination. This was the sort of device that a number of people, including Galileo himself, had been working on in the previous decades. Receiving more detailed information from Jacques Badovere, an acquaintance in Paris, Galileo realized that Lipperhey's spyglass was entirely lens-based. This gave him enough clues to work out how it worked and how to make one from a lead tube and two glass lenses. Galileo's first instrument magnified only about three times, but he found that he 'perceived objects satisfactorily large and near' with it. Over the next few months he worked out how to improve the design and build more powerful examples.

Although he was by no means the first to make or use a telescope, Galileo was very quick to exploit the potential of what he realized could be a powerful military tool. He also knew it could make his fortune – and a fortune was what he needed to support his lifestyle. As early as August 1609, he demonstrated an eight-powered version to the Venetian Senate, persuading the elderly senators to climb up the Campanile di San Marco to see it used to spy out enemy ships. He emphasized the strategic value again in a letter to the senators, pointing out that the spyglass was 'of inestimable benefit for all transactions and undertakings, maritime or terrestrial'. It allowed one to spot hostile ships in time to prepare a defence, or look into fortresses to watch the enemy's preparations.

Within a few days, the Senate offered him lifetime tenure at the University of Padua on the understanding that they would retain the right to manufacture the instrument. The offer doubled his salary, making him one of the ten highest earners in Tuscany, but Galileo was already seeking a better appointment at the Medici court in Florence. He succeeded by turning his spyglass to the heavens. He also conveniently ignored the question of production rights.

Galileo was not the first to do even this. Early reports of Lipperhey's device already talked of viewing the stars. And in London, Thomas Harriot – an employee of the Earl of Northumberland who had been briefly imprisoned after the Gunpowder Plot of 1605 – recorded observations of the Moon with his six-powered 'trunk' on 26 July 1609.

Galileo's astronomical observations began a few months later in the autumn of 1609. He may have been lagging behind other observers at this stage, but almost immediately he saw remarkable things with his twenty-powered *perspicillum*. They were so remarkable that he rushed his findings into print, published the following March in the *Sidereus Nuncius*, the 'Starry Message' (or 'Starry Messenger').

Sidereus Nuncius was a modestly sized book but its astonishing revelations were obvious and quickly made him famous. Put simply, Galileo claimed to have discovered that the Moon was not a smooth sphere but had mountains and valleys; that the Milky Way was made of individual stars; that there were more stars in the heavens than the unaided eye could see; and that Jupiter had four moons circling around it. These were radical claims that flew in the face of the established doctrine that the Moon and planets were perfect and unchanging, and held in spheres revolving around the Earth. As a supporter commented, it was a 'bold deed, to have penetrated the adamantine ramparts of heaven with such frail aid of crystal'.

It was the moons of Jupiter that proved to be the trump card. By naming them the Medicean Stars, Galileo was able to present

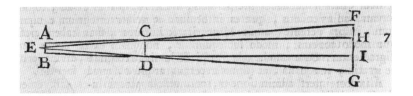

8   Galileo's sketch of the workings of his *perspicillum* in the *Sidereus Nuncius*, from Galileo Galilei, *Opere* (Padua, 1744). The diagram deliberately lacked details of the instrument's construction and dimensions, preventing potential rivals from building their own.

the Grand Duke of Tuscany with a gift that would immortalize his family's name. In return for this and dedication of the book to the Duke, Galileo requested a court position. And his wish was realized in July 1610 with the offer of a new post as Chief Mathematician of the University of Pisa and Philosopher and Mathematician to the Grand Duke of Tuscany. For a second time, he successfully used the telescope to further his career.

Still Galileo felt he could gain more from 'his' invention and its future discoveries. But to do so, he needed to control the use of his *perspicillum*. The *Sidereus Nuncius* was therefore deliberately vague about how to make it, with a sparse description accompanied by a basic illustration (Fig. 8). There was little detail for anyone wishing to build their own and repeat or add to Galileo's discoveries.

Galileo's other tactic was to dictate who owned the instruments that he made and distributed around Europe. Using the Medici's ambassadors as messengers, he sent them as gifts only to the most politically powerful – princes, cardinals and ambassadors – rather than to mathematicians or philosophers with a more academic interest. So Johannes Kepler never managed to get a telescope from Galileo despite repeated pleading, although this did not prevent him from writing a glowing response to the *Sidereus Nuncius*. Eventually he looked through one that was given to the Archbishop-Elector of Cologne.

This selective giving also gained Galileo valuable credit with potential patrons, for it was not money he sought but personal advancement. So when Philip IV of Spain asked to purchase one of the famous spyglasses, Galileo responded that he never sold his instruments and never would.

Galileo also organized personal viewings throughout Italy. Those invited were taught to look through the instruments properly – to identify the satellites of Jupiter, say – but were not shown how the apparatus was made. Some found ways around this secrecy: in April 1610, an unscrupulous student named Martin Horky surreptitiously made a wax cast. Boasting about it in a letter to Kepler, Horky said that it was his intention to make a better instrument than Galileo's, but his device never materialized.

Lastly, Galileo sought to ensure that others only used spyglasses he deemed to be of sufficient quality, by which he meant those able to repeat his own observations. This was crucial. In the early years, telescopes were of modest capability. By its nature, the Galilean telescope has a limited field of view, allowing Harriot and Galileo to see only a quarter of the Moon at a time. On top of that, the lenses suffered from distortions and aberration, producing blurred images with distracting coloured rings around their edges. Other factors made successful observation difficult. Galileo found that even breathing caused his hand to shake and so recommended fixing the tube 'in some stable place'. The lenses also needed to be wiped regularly with a cloth to prevent the observer's breath or other vapours clouding them over.

It was difficult to make successful observations with a telescope of the best quality available, and many of those in use were not the best. Many times Galileo grumbled about the difficulties he faced in producing good instruments. In March 1610 he complained that, 'spyglasses that are most exquisite and capable of showing all the observations are very rare, and among the sixty I have made, at great cost and effort, I have been able to find only a very small number'. The main problem was obtaining good lenses. At a time before there were many dedicated manufacturers of optical instruments, Galileo had to commission work from other makers with the appropriate skills – mirror-makers, *pietra dura* polishers or spectacle-makers. The quality of glass was also variable. Glass for mirrors was generally the best available, so old or broken mirrors were often re-ground into lenses. In April 1616 Giovanfrancesco Sagredo reported to Galileo that he had twenty-two lenses from a mirror-maker in Murano. These were the best of 300 produced in the previous weeks; only three were remotely suitable and were still not good enough for Galileo's purposes.

The varying quality of lenses and of the instruments made with them meant that some observers were unable to see the things Galileo wrote about. Unsurprisingly, they doubted his claims. Jacob Christmann complained that the appearance of Saturn depended on which telescope he used. He concluded that the different telescopic appearances must be little more than fantasy. Galileo himself initially thought that Saturn consisted of three bodies in a line; the companion bodies seemed to have disappeared at a later date. Others believed it was oval-shaped. It was only in 1655 that Christiaan Huygens described the planet's now familiar rings, resolved with the aid of a more powerful telescope than those available forty years earlier.

A significant mental obstacle lay at the heart of these early disagreements. The telescope was the first instrument to extend one of the human senses. It was not a given that what it showed could be accepted without question. Having passed through pieces of glass, could the images be guaranteed as real, rather than artificial products of the instrument? Horky's *Brevissima peregrinatio contra nuncium sidereum* (Brief Foray against the Starry Messenger, 1610), the first attack on Galileo's discoveries, lambasted his claims on precisely these grounds. Galileo's longtime enemy, Lodovico delle Colombe, drew an analogy from painting to suggest that the apparently rough lunar surface Galileo described was merely an optical illusion. The most outrageous claim came from the Jesuit mathematician Christopher Clavius, who in 1610 reportedly said that Galileo could not have seen Jupiter's moons unless he had first put them into the telescope (although Clavius did confirm Galileo's observations a year later).

Hoping to answer his critics, Galileo continued to demonstrate his instrument in person, with varying success. In April 1610, after a visit to Giovanni Antonio Magini, Professor of Mathematics at the University of Bologna, Horky wrote to Kepler that, 'On the Earth it works miracles; in the heavens it deceives'. Similar resistance was evident a year later when Galileo demonstrated the telescope at a villa outside Rome. First he sought to show that the telescope did not alter what was seen by apparently encouraging his audience to read the inscriptions painted in the gallery of the nearby Lateran Palace. Having assured them that the telescope revealed only what was really there on Earth, he suggested that what was seen in the heavens was just as real. Yet one of those present, Giulio Cesare Lagalla, Professor of Logic and Philosophy at the University of Rome, refused to accept the celestial observations, even though he acknowledged the reality of the terrestrial ones. Like Colombe, Lagalla argued that Galileo's lunar mountains and craters were an optical illusion. According to the long-accepted Aristotelian view of the universe, the terrestrial and celestial regions were made of completely different matter. It did not follow that the telescope worked equally faithfully in both regions.

Nonetheless, the matter was largely being put to rest in Galileo's favour at about this time. The turning point came with official support from the Collegio Romano, the Jesuit training college in Rome. Cardinal Bellarmine, head of the college, wrote to his mathematicians in March 1611 to ask their opinion of Galileo's observations. Just five days later Fathers Clavius, Grienberger, Lembo and Maelcote replied, confirming that the observations were real, but disputing Galileo's reading of them.

The Jesuit response shifted the debate from the reality of telescopic observations to their interpretation, particularly Galileo's religiously suspect claim that the Moon was irregular. Sure enough, the Moon did not look smooth and uniform even to the naked eye, but this did not prove that it was not a perfect sphere. The Jesuit Fathers suggested instead that its appearance was due to its non-uniform density rather than any surface irregularity. Others likened it to a crystal ball or to a sunlit cloud, explaining the apparent imperfections as tricks of the light.

Debates continued as further discoveries were made in the years after the publication of *Sidereus Nuncius*. Towards the end of 1610, Galileo announced that Venus exhibited phases similar to those of the Moon. Since this could be interpreted as demonstrating that the planet orbited around the Sun, it was crucial support for the heliocentric system. Those wishing to retain the geocentric model opposed this conclusion.

Two years later, Galileo was embroiled in an argument with another Jesuit mathematician, Christoph Scheiner, over the discovery and significance of sunspots. Scheiner argued that these were bodies lying between the Sun and the Earth, while Galileo held that they were on the Sun's surface; again, contradicting the assumed immutability of the celestial bodies. In his response, Scheiner did not deny the reality of the phases of Venus or sunspots as observed; he argued for their interpretation within a geocentric model. Ironically, this was not a one-sided process. Galileo also had to admit that some appearances were not as they seemed, most obviously the visibly smooth edge of the Moon, which should have appeared rough and mountainous according to his own claims.

Overlooking these differences of interpretation, the mathematicians of the Collegio Romano honoured Galileo for his achievements when he visited Rome in 1611. The many other festivities during his visit included a feast on 14 April, at which Galileo was made a member of the Accademia dei Lyncei (Academy of the Lynx-eyed), a select scientific society. The celebration was doubly notable, for during the feast Federico Cesi, the Academy's founder, unveiled a new name for the miraculous instrument, suggested by

9   Galilean telescope, inscribed with the name of Jacob Cunigham, 1661.

the Greek theologian John Demisiani. The name was *telescopium*, derived from the Greek τηλεσκόλος (*telescopos*), meaning 'far-looker'. This settled the question of what to call an instrument that had already gone by many other titles – *perspicillum, organum, instrumentum, occhiale, Batavica dioptra*, to name just a few. The telescope had arrived.

While Galileo wished to control the manufacture and ownership of new telescopes, realistically he could never achieve it. Already by the end of 1609, low-powered telescopes had changed from exclusive devices to cheap gadgets sold by travelling artisans and local spectacle-makers. By February 1610 they were even being made in London.

In these early years all telescopes, whether they were cheap novelties sold on the streets of European cities or Galileo's finely crafted gifts, had the same basic form, a convex and a concave lens held in a tube. This is the form now called the Galilean telescope (Fig. 9). It is also important to remember that most of these telescopes were for terrestrial uses, in spite of the groundbreaking astronomical work of Galileo and his contemporaries. The new optical aid was valuable in war for revealing the enemy's manoeuvres, the English admiral Sir William Monson lamenting the fact that many strategies would have been more effective had 'prospect glasses not been so common'. It had other more peaceful uses on land. Sagredo wrote to Galileo in 1618 that spyglasses with short tubes were on sale in Venice and that he had used one for examining paintings. Uses were found in any situation where people wanted to see things closer up.

Collapsible hand-held telescopes with nested draw tubes were also introduced within a few years. They were depicted as early as 1614 and were being produced commercially by 1617, with Hieronymous Sirturus in Venice the first to describe them in detail. It was Sirturus who pioneered the use of tubes made of cardboard or pasteboard, which were light and rigid and needed only a suitable covering for protection. It was a form that remained standard for 165 years or more and was still being used 200 years later. Before then, the tubes were made of wood or metal.

The telescope was also spreading beyond Europe. An early example from South America is found in a report of the Battle of Guaxanduba on 19 November 1614, where Jeronimo de Albuquerque, commanding Portuguese forces on the coast of Brazil, was attempting to oust French colonists on the island of Saõ Luiz. According to an official observer, the commander was at one point seen peering at the enemy with 'a glass for seeing at a distance'. The observer tartly commented that, 'this is not the time to be looking through telescopes, for it will neither lessen our task nor make our enemies fewer'.

The first telescope to arrive in Japan is said to have been a gift to the local ruler of Hirado from John Saris, captain of the *Clove*, an East India Company vessel sent to develop trade. By contrast, Jesuit missionaries brought the telescope to China in an astronomical context. Aiming to introduce Christianity there, one of the Jesuits' methods was to demonstrate the sophistication of European mathematics and astronomy. As early as 1615, Emmanuel Diaz produced a Chinese account of Galileo's discoveries. Eight years later a party of Jesuit missionaries arrived in Peking with gifts including a telescope. Among them was Johann Adam Schall von Bell, who promoted the use of the telescope in astronomy and wrote the first book in Chinese on the instrument.

The initial take-up in China was limited. Chinese astronomers were mainly concerned with calculating and predicting eclipses and the movements of the heavenly bodies to produce precise calendars, which underpinned the Emperor's mandate as ruler. Investigating the nature of the heavenly bodies was not their main interest, although they appreciated the telescope's use for revealing the phases of Venus and the motions of Jupiter's moons.

Although the telescope did not answer the immediate needs of Chinese astronomers, optical devices had a long tradition there, with burning lenses probably known in the first century AD and optical lenses by the tenth century. This tradition has led to some speculation as to whether telescopes were ever invented in China, but there is evidence that their origin was purely European in a digression in Li Yu's short story, 'A Tower for the Summer Heat' (1657). In the tale, a young scholar pursues marriage to a beautiful woman by using the far-seeing potential of a 'thousand *li* glass' to persuade her that he has supernatural powers. Li Yu's description of the scholar's instrument notes that telescopes first came from Europe as gifts to the Chinese emperor.

Back in Europe, the telescope was undergoing further development. It stemmed from the description of a new form of telescope in Johannes Kepler's *Dioptrice* (1611), a work on the theory of lens systems. The Keplerian or astronomical telescope comprised two convex lenses and had the advantage of a larger field of view than the Galilean. One or two astronomers soon experimented with it, notably Christoph Scheiner as early as 1617 (Fig. 10). These rare exceptions aside, the idea was largely overlooked for a number of years because the Keplerian telescope produced an inverted image when used for direct observation and was no use on land or at sea.

A generation later, a work entitled *Oculus Enoch et Eliae* (1645) marked a more significant turning point, after which hand-held

10   Observing the Sun with an astronomical telescope, from Christoph Scheiner, *Rosa Ursina* (Bracciano, 1630). Scheiner used Kepler's telescope design to project an image onto a screen, so that the telescope's inverted image appeared the right way up. It also protected Scheiner's eyes since he did not have to look directly at the Sun.

and astronomical telescopes began to develop along different paths. The book was by Johann Burchard Schyrle, who had taken the name Anton Maria Schyrle de Rheita on becoming a Capuchin monk. From 1645 Schyrle was carrying out scientific work in Cologne and Augsburg with the lens-grinder Johann Wiesel. His book describes in a coded way some improvements to the Keplerian telescope, introducing what is now called the terrestrial telescope. The modified design included additional lenses between the two convex lenses of Kepler's proposal (see Glossary). These increased the field of view and re-inverted the image so that the observer saw it the right way up. This was a telescope that was ideal for terrestrial purposes. For astronomy the added lenses caused too much loss of light, but using Kepler's original two-lens system reduced this problem. From about 1645 astronomers like Christiaan Huygens began using Keplerian telescopes much more widely. They were not particularly troubled by the inverted images the telescopes produced.

11 Portable telescope by William Longland, London, *c.* 1690. Longland was one of the English makers who adopted Wiesel's designs.

Wiesel was selling the new terrestrial telescopes by 1647. 'With these', his advertisement announced, 'an entire army of about 7 or 8 thousand may be beheld clearly at one time and be examined very distinctly indeed.' It was not long before telescopes of a similar design were also being produced by makers elsewhere in Germany, France, Italy and England (Fig. 11). They were the most common form for terrestrial telescopes for the next hundred years and were popular into the nineteenth century.

The success of the terrestrial telescope meant that the Galilean configuration was largely abandoned, except for low-power perspective or opera glasses, which were useful for the short-sighted. A letter of 1656 reveals that the myopic King Gustavus Adolphus of Sweden 'used a short perspective so concealed in his hand that it was not discerned: whilst he seemed to regulate the brim of his beaver, he used this perspective'. Similar devices were also used by hunters and on the battlefield, and became fashion accessories at the theatre and other social events. On a Sunday in May 1667, Samuel Pepys noted, 'I did entertain myself with my perspective glass up and down the church, by which I had the great pleasure of seeing and gazing at a great many very fine women.'

Aside from the development of the terrestrial telescope, until about 1660 the major technical advances were largely in lens-grinding techniques, with those who mastered them proving to be in great demand. It is easy to imagine the frustration of the French king, Louis XIV, who was forced to make lucrative offers for the secrets of two leading Italian lens-makers, Eustachio Divini and Giuseppe Campani, only for both to refuse. It was an art that called for great secrecy. Evangelista Torricelli (Galileo's successor at the Medici court) wisely insisted on his deathbed that his lens-making secrets and equipment be locked in a crate and kept for the use of the Grand Duke alone.

Despite advances in lens production, makers and users of telescopes realized that the optical defects that lenses sustained could not be readily eliminated. Instead, the best way of reducing image distortions was to grind lenses with decreased curvatures and to restrict their aperture, since the outer parts of lenses tended to be more imperfect. As the use of telescopes for astronomy took off from the mid-1640s, their lengths steadily increased to accommodate these lenses which had longer focal lengths. By the mid-1650s, the astronomical telescopes made by Christiaan Huygens were about 23 feet long. Twenty years later, some were twice this length.

Telescopes for astronomical purposes now became too large to be handled by a single person and too expensive for private individuals. One exception was Johannes Hevelius, whose wealth as the son of a rich brewing family was supplemented by his marriage to Katharine Rebeschke. Katharine, the daughter of a wealthy citizen of Danzig (Gdańsk), lived next door to Hevelius and owned the two neighbouring houses. This enabled him to follow his interest in astronomy by building large telescopes and an observatory that spread across the rooftops. From there he made detailed maps of the Moon that were the best available for a century.

Hevelius and his second wife Elisabeth then worked towards a new star catalogue, *Prodromus astronomiae* (1690). What is surprising is that he chose not to use telescopic instruments for his measurements but large naked-eye instruments like Tycho Brahe's of the century before. This put him at odds with recent developments in Europe. Telescopes used as sighting aids for astronomical sextants were first described as early as 1611, but it was as the astronomical telescope was being developed in the 1630s that this took off, with Jean-Baptiste Morin suggesting the use of telescopic sights again in 1634. Five years later, the English astronomer William Gascoigne created more accurate sights by stretching threads from a spider's web inside his telescope, later using human hair for the same purpose. Gascoigne's death at the Battle of Marston Moor in 1644 during the English Civil War prevented the idea from being more widely known, but by the 1650s others had devised similar means of incorporating sighting lines and measuring devices into telescopes. These included Christiaan Huygens's published account of an eyepiece micrometer in 1657. These innovations allowed telescopes to be used for astronomical measurements in two ways: as sighting aids on instruments such as quadrants, used for measuring large angles between different celestial bodies; or to measure small angles such as a planet's diameter by means of a micrometer.

While Hevelius was happy to use telescopes for observation and mapping, he did not believe that instruments with telescopic sights could perform better than naked-eye instruments for angular measurement. Among other things, he complained that the observer's breath clouded the lenses. At the Royal Society in London, William Molyneux dismissively wondered, 'how easily is this avoided? Who is it goes purposely to make a Speaking-Trumpet of a Telescope?' Meanwhile, Robert Hooke began an increasingly heated argument with Hevelius. To settle the matter, an up and coming astronomer named Edmond Halley was sent to Danzig in 1679. He found, to everyone's surprise, that Hevelius's measurements were as accurate as any made with telescopic sights. But Hevelius and his instruments were exceptional. Elsewhere, the use of telescopic sights did increase accuracy. Telescopically aided observations made at the end of the seventeenth century by John Flamsteed, the

12 The aerial telescope, from Christiaan Huygens, *Opera Varia, Vol . I* (Lugduni, 1724). Huygens' design did away with the long, unwieldy barrel. The objective lens is mounted on a pole, while the observer holds the eyepiece.

first Astronomer Royal at Greenwich, proved to be about twenty times as accurate as Tycho Brahe's one hundred years earlier.

As the debate over the use of telescopic sights showed, the nature and use of the telescope was still far from agreed in the late seventeenth century. This was also true of the ways in which large astronomical telescopes were evolving. As they grew increasingly long and cumbersome, astronomers began to look for innovative solutions to the problems they presented.

By the 1670s some of these monster telescopes had grown to extraordinary lengths, including one of over 140 ft (43 m) that Hevelius had erected on the beach at Danzig. A telescope like this had to be supported from a mast and raised and lowered with ropes and pulleys by many assistants, yet proved frustrating to use as a slight wind would move the framework (see also Fig. 14). One solution came from Christiaan Huygens, who suggested removing the tube completely (Fig. 12).

Other astronomers proposed different forms of tubeless telescope. In 1669 Robert Hooke created one by cutting a hole from the ground floor up through the roof at his lodgings at Gresham College, London. It proved too unstable for accurate measurement, but the idea of an architectural solution was one he pursued. Following the Great Fire of London, the commemorative Monument designed by Hooke and Christopher Wren, which was erected next to the Thames in 1671–77, was also a giant vertical telescope. It too failed to produce the desired results: the structure swayed in the wind and expanded and contracted as the temperature changed.

Another idea was to build a telescope not upwards but downwards. John Flamsteed used a well at the Royal Observatory in Greenwich for this purpose, hoping that it would give him a stable 'tube' into which he could build a long-focus telescope (Fig. 13). Like the Monument and Hooke's telescope at Gresham College, it was intended to measure stellar parallax by observing one of the stars that passed directly overhead. Flamsteed made some observations from the well in the summer of 1679, but the experiment failed 'because of the damp of the place' and it was subsequently filled in.

While the problem of increasing length led to novel suggestions, size alone did not guarantee quality. Much of the productive

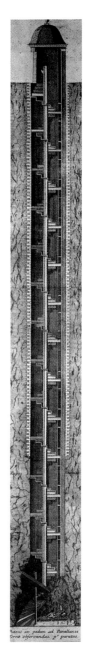

13 John Flamsteed's well telescope at the Royal Observatory, Greenwich, engraved by Francis Place from a drawing by Robert Thacker, c. 1676. The observer lay on a couch at the bottom of a shaft that was 100–130-ft (30–40-m) deep and observed the stars directly overhead.

14 *Greenwich Royal Observatory from Crooms Hill*, by an unknown English artist, c. 1690. At the rear of the Observatory is a 60-ft (18.3-m) telescope that hung from an 80-ft (24.4-m) mast. Only one observation was recorded with it. It began to sway dangerously in 1690 and was removed by the end of the century.

practical astronomy of the period was done with telescopes of more modest lengths, including Hooke's measurement of the period of Mars's rotation with a telescope of 36 ft (11 m). In the end, their bigger cousins were more useful as public spectacle than as accurate scientific instruments, although not all were seen as failures. The Royal Society regarded a 60-ft (18.3-m) telescope used by Hooke as emblematic and pictured it on the title page of Thomas Sprat's account of their early years as one of the wonders of a new age of reason.

The other significant development was that of the astronomical observatory. The increasing size of telescopes meant that they needed not only dedicated space but also considerable funding. Rich exceptions like Hevelius aside, this meant state support. The first state observatory was founded in Copenhagen in 1637, and in subsequent decades other countries followed suit. These included national observatories founded in Paris in 1667 and Greenwich in 1675 (Fig. 14), each set up to solve the navigational problem of determining longitude through the production of accurate star tables. Both were also at the forefront of developments in the construction and use of astronomical telescopes. In Paris, huge advances in their use on angle-measuring instruments were made under Giovanni Cassini, the first of four generations of the same family to head the observatory. At Greenwich, John Flamsteed promoted telescopes for star mapping and carried out important studies into the theory and proper use of telescopes in astronomy.

By the end of the century, the telescope had evolved from its modest beginnings to become an important seeing aid. On land and at sea, it was a common tool for military and civilian users. Pointed to the skies, it was becoming an indispensable astronomical instrument that was calling into question whether the sky was indeed the limit.

Chapter Three

THE TELESCOPE AND THE IMAGINATION 1608–1700

Oh, by a trunk! I know it, a thing no bigger than a flutecase: a neighbor of mine, a spectacle-maker, has drawn the moon through it at the bore of a whistle.

Ben Jonson, *News from the New World Discovered in the Moon* (1620)

While the telescope had a big impact in practical terms as a naval and military tool and in astronomy, its influence extended deeper into seventeenth-century culture. It clearly had a link to those arts connected with seeing. It also affected other creative pursuits, which responded not only to the possibilities presented by a new instrument for extending one of the senses, but also to its revelations about the universe.

In the visual arts, the possibilities offered by the telescope found fertile ground within recent trends, especially the growth of landscape painting and other genres that promoted a new naturalism in art. The telescope's English names included 'prospect glass', 'perspective glass' and 'perspective', and so the device became entwined with the representation of landscape as the terms 'landscape', 'prospect' and 'perspective' became virtually synonymous. The latter terms encompassed both the panoramic view and the optical device that could produce it.

In Italy, artists including Galileo's friend Ludovico Cigoli began using telescopes shortly after they appeared, and for a brief period painting the 'telescopic Moon' became a common artistic practice in Tuscany. Within a few months of the publication of the *Sidereus Nuncius*, Cigoli incorporated Galileo's lunar discoveries into his painting for the cupola of the Pauline (or Borghese) Chapel in the Basilica of Santa Maria Maggiore, Rome. This showed the Virgin Mary standing on an imperfect Moon with a rough surface, just as Galileo described. Adam Elsheimer's *The Rest on the Flight into Egypt* (Fig. 15) showed a spotted Moon and the individual stars of the Milky Way that Galileo's telescope revealed. Although these proved to be early exceptions, since for some time other paintings of the Moon continued to represent it as smooth and perfect in line with religious doctrine, a naturalistic tendency in the representation of the heavens did gradually evolve in tandem with the telescope's new discoveries.

Telescopes also began to appear as subject matter in paintings. As a key instrument of navigation and naval warfare, they became

15 *The Rest on the Flight into Egypt*, by Adam Elsheimer, 1609. Elsheimer's seemingly naturalistic rendering includes a rugged Moon and individual stars in the Milky Way, as if inspired by news of Galileo's discoveries. It is not an observable sky, but combines astronomical features symbolically linked to the Virgin Mary and Child.

a natural accessory with which a captain or admiral might be painted, an instantly recognizable attribute of their profession (Fig. 16). For similar reasons, they also began to appear among the accoutrements of astronomers and other scientists.

Other representations placed the telescope in more symbolic settings, notably the moralizing allegories of the senses that were already popular at the beginning of the century. *Sight* (1617) by Brueghel and Rubens, shows a female figure representing that sense surrounded by a bewildering array of instruments, creatures and images, each signifying a different aspect of seeing or optics (Fig. 17). The telescope placed prominently in front of her appropriately records the age's most recent seeing instrument.

16 *Sir Cloudesley Shovell* (1650–1707), by Michael Dahl, 1702–05. Shovell was one of the English navy's most popular commanders. The prominently displayed telescope indicates his profession, but also symbolically replaces the baton of command of earlier military portraits.

More generally, an important effect of the new telescopic discoveries was to transform how people thought about the universe around them and their place within it. By challenging the established vision of a finite universe comprised of solid spheres, telescopic revelations began to present the cosmos in a fresh light. Several authors likened the latest astronomical discoveries to those made on the world's oceans a century or more earlier, which had been just as startling. Galileo would now 'compete in glory with Columbus', said Orazio del Monte, a judgement echoed by Sir William Lower, for whom Galileo had 'done more in his threefold discoverie than Magellane in opening the streights to the South Sea'.

Just as Columbus's discoveries expanded European concepts of the extent of the Earth and its inhabited regions, so telescopic

17 *Sight*, by Jan Brueghel 'the Elder' and Peter Paul Rubens, 1617.

discoveries opened up the possibility of vast expanses of space that might contain other life. These were powerful notions that authors readily took on board, seriously and satirically.

The work of the metaphysical poet and preacher John Donne followed both lines. In an anti-Catholic polemic, *Ignatius his Conclave* (1611), he imagined Galileo creating a new form of telescope that could pull the Moon to the Earth as a vehicle for expelling the Jesuits he despised. Yet not long afterwards Donne was taking the implications of telescopic observations more earnestly and thinking about humanity's place in a larger, changeable universe. In *An Anatomy of the World* (1611), a more sober Donne reflects:

> And new philosophy calls all in doubt,
> The element of fire is quite put out;
> The sun is lost, and th'earth, and no man's wit
> Can well direct him to look for it.

Similar thoughts are to be found in the poetry of John Milton, who even claimed to have visited Galileo in Italy. His epic poem *Paradise*

Lost (1667) is filled with a sense of the immensity of space that the telescope was beginning to reveal, with Earth but 'a spot, a grain, / An atom' in comparison to Galileo's newly discovered universe. Milton's description was probably the first attempt in English poetry to picture in detail this new limitless cosmos, and explicitly acknowledges the impact of astronomy's newest instrument, the telescope, describing Satan's shield, of which,

> ... the broad circumference
> Hung on his shoulders like the moon, whose orb
> Through optick glass the Tuscan artist views ...
> Book I

In the works of Milton and others, the telescope and its discoveries accompanied a new realism in astronomical description and an awareness of the expanded universe. This included renewed speculations about the possibility of life on other planets. As early as 1610, Kepler discussed this in his *Dissertatio cum Nuncio Sidereo* (Conversation with Galileo's Sidereal Messenger), concluding not only that the Moon has inhabitants, but that Jupiter does as well. He argues that its four satellites can only have been created for the Jovian peoples, just as Earth's moon was created for humanity.

Ben Jonson's *News from the New World Discovered in the Moon* (1620) took up the same theme, with reports of the lunar lands and its inhabitants brought back to Earth by 'moon-shine' collected 'by a trunk' (a telescope). Two decades later, Francis Godwin's fictional *The Man in the Moone* (1638) told of a journey to an inhabited Moon, while in the same year John Wilkins's *The Discovery of the World in the Moon* speculated quite seriously about the possibility of life there. Both credited Galileo and his telescope as their inspiration, Wilkins referring to 'Galilaeus, whose glasse can shew this truth to the senses'. Robert Hooke's famous *Micrographia* (1665), which revealed wonderful discoveries with that other optical novelty, the microscope, also considered that,

> 'Tis not unlikely, but that there may yet be invented several other helps for the eye . . . by which we may perhaps be able to discover *living Creatures* in the Moon, or other Planets . . .

The telescope and its discoveries forced people to think again about the universe and its operations in other ways too. This was obvious to Sir Henry Wotton in 1610, who pointed out that as well as questioning current thinking about the physical nature of the cosmos, Galileo's discoveries challenged the rules of astrology, which was still widely accepted. If there were new stars and planets, he reasoned, should their influence not also be included in any astrological forecast? It was an argument that would be repeated many times, but like the question of whether the heavens were perfect and unchangeable, the evidence was not conclusive. Kepler's *Dissertatio* also questioned whether Galileo's discoveries threatened astrology but decided that they did not.

For many, the telescope was evidence of humanity's tremendous advancement, a wondrous device that emphatically proved the ingenuity of the 'moderns' and their superiority over the 'ancients'. Henry Power wrote in his *Experimental Philosophy* (1664) that, 'Dioptrical Glasses are but a Modern Invention. Antiquity gives us not the least hint of them'. The ancients, he argued, produced nothing like the microscope or the telescope, which revealed new worlds and prophesied new universes. Robert Hooke went further: the new instruments might allow humanity to regain some of the perfection lost in the Fall. Others were more doubtful, believing that the 'moderns' were still very far behind. Even Jacob Metius, one of the first to claim the telescope's invention, only asserted that he had been investigating, 'some hidden knowledge which may have been attained by certain ancients through the use of glass'.

With the telescope's introduction during a period of widespread religious turmoil in Europe, it is not surprising that many writers used it for moral and religious ends. As an instrument that revealed things previously invisible to the human eye, what could prevent it from metaphorically uncovering spiritual truths, as in John Vicars's

18   Francis Quarles, *Emblemes* (1635), Book 3, emblem XIV.

A Prospective Glasse to look into Heaven (1618)? Pictorial uses of the telescope also took on this moral, clairvoyant quality. In Francis Quarles's *Emblemes* (1635), an illustrated book of religious meditative verse, a clothed spirit uses a telescope to view death and the flames of hell beneath angels in heaven (Fig. 18). The image takes its moral from the book of Deuteronomy (32:29), 'O that they were wise, then they would understand this; they would consider their latter end'. It is followed by a discussion between Flesh and Spirit in which the telescope brings these possible destinies into focus, allowing a clearer contemplation of one's future actions. Milton uses the same idea in his *Prolusiones* (1674): 'And in times long and dark prospective Glass / Foresaw what future days should bring to pass'. The foreknowledge the telescope offered was best used to put one onto a proper path.

Other artistic and literary uses were less reverend. In about 1630, Niccolò Aggiunti wrote a poem about two gifts from his friend Galileo: a telescope and some Greek wine. The optical instrument enlarged images, Aggiunti wrote, the wine multiplied them; two glasses (the lenses) took him to the stars, but four glasses (of wine) transported him above them; and while the telescope allowed him to see Venus's horns (the planet's phases), the alcohol revealed how she gave them to Vulcan (referring to the horns of the betrayed husband).

Aggiunti's playful poem was to be taken in good spirit (in more ways than one), and it did not take long for telescopes to become the objects of humour. As early as 1615 Thomas Tomkis's *Albumazar* has the astrologer's sidekick trying to impress an intended dupe with the miracles his master can perform by showing off his 'perspicill', an 'engine to catch starres'. Tomkis was the first writer to anglicize the Latin word Galileo used for the new instrument in the *Sidereus Nuncius*.

By the second half of the seventeenth century, the telescope's comedic appearances were bound up with swipes at virtuosi, those dabblers in scientific speculation. Virtuosi became stock figures in satirical plays featuring foolish, would-be scientists oblivious to the real world around them (the predecessors of the absent-minded professor). As a stage prop, a telescope added to the ridicule, recalling the ever longer examples audiences might have seen in real life. Theatrical telescopes would be deliberately long and cumbersome: the stage directions in Aphra Behn's *The Emperor of the Moon* (1687) call for a telescope of 20 ft (6 m) or longer. Those who used such unwieldy instruments, audiences understood, were gullible and prey to wild speculations. The telescope also emphasized the virtuoso's ignorance of what was right under his nose, since its use distorted vision and involved looking away from the Earth. The appropriately short-sighted Doctor Baliardo of *The Emperor of the Moon* is so caught up in his nightly observations that he is easily persuaded by his daughter's lover that the Moon's ruler has come down to take her as a bride.

Taking this further, Samuel Butler's satirical poem *The Elephant in the Moon* (1676) ridicules his virtuosi as not just stupid but morally lacking. His telescope symbolizes their distorted vision: they are interested only in a hypothetical lunar elephant, not an actual earthly mouse. Observing the Moon through their 'optic engine', the group believe they can see a battle with an elephant caught up in it, leading them to wild speculations about lunar life until a servant uncovers the truth:

> And, viewing well, discovered more
> Than all the learned had done before.
> Quoth he, 'A little thing is slunk
> Into the long star-gazing trunk;
> And now is gotten down so nigh,
> I have him just against mine eye.

Butler's virtuosi cannot accept this possibility until the telescope is opened up and the battling figures are found to be flies and gnats trapped with the mouse they have mistaken for an elephant. Instruments like the telescope, Butler implies, cannot help those who refuse to see the truth because of their moral blindness. The attack was all the more scurrilous because the poem's targets were recognizable as members of the Royal Society, including Sir Paul Neal, John Evelyn, Robert Hooke and Robert Boyle. Butler charged them with seeking marvels rather than truth.

The same theme occurs in Thomas Shadwell's *The Virtuoso* (1676), in which Sir Nicholas Gimcrack's laughable investigations include observing similar lunar battles. Emphasizing that their work has no practical purpose, Gimcrack even confesses that, 'virtuosos never find out anything of use, 'tis not our way'. This was a key criticism of those dabbling in philosophical and scientific pursuits, as Margaret Cavendish's *The Description of a New World, Called the Blazing World* (1666) also makes clear. Cavendish criticizes her contemporaries' preoccupation with building instruments for their own sake rather than attempting to develop something useful to society.

In her imagined world, she lampoons the frustrating arguments between the bird-men (who are astronomers) about what they see through their telescopes. Angered by their never ending disputes, the Empress finally cracks and cries out that, 'your glasses are false informers, and instead of discovering the truth, delude your senses; wherefore I command you to break them'.

In highlighting again the problem of deciding what a telescope has shown and what this might mean, Cavendish repeated a problem that surfaced in the first uses of the instrument at the beginning of the seventeenth century. The same question would crop up again and again long after the telescope was established as a key tool of science. Yet in its first century, the telescope had an undeniable impact on the popular imagination, establishing itself as a source of positive imagery and speculation, an icon of scientific technology, and a victim of satirical humour.

Chapter Four

THE REFLECTING TELESCOPE 1610–1800

I am still of the opinion that I have seen the two most wonderful things that have ever been seen in this Planet: the French Revolution and Dr Herschell's telescope.

John Anderson, letter to James Lind, 7 October 1791

The telescope Hans Lipperhey unveiled in 1608 used lenses to extend human vision, but many people were already considering making far-seeing devices with mirrors. The introduction of the Dutch telescope encouraged mathematicians and philosophers to think again about using mirrors in a similar way. But it would take more than a century before these instruments became widely available. Then in the later part of the eighteenth century a family who came to England from Europe was to have a profound impact on the development of reflecting telescopes for astronomy.

The first claims to have built working telescopes with mirrors surfaced within a few years of Lipperhey's announcement. These included the Italians Niccolò Zucchi, who said he had made one as early as 1616, and Cesare Caravaggi, about whom Galileo heard similar news ten years later. Bonaventura Cavalieri, a professor at the University of Bologna, and the French mathematician Marin Mersenne also wrote about telescopes based on mirrors, but did not successfully build them. The same was true of a Scottish mathematician, James Gregory, whose *Optica Promota* (1663) described several telescope designs, including an arrangement still known as a Gregorian reflector (see Glossary). Gregory's attempts to produce a working example proved to be beyond the abilities of even the best London makers, including Richard Reeve, the lens-maker favoured by Samuel Pepys.

The principle of the reflecting telescope is simple. It uses a curved mirror instead of a lens to collect light, the only problem being that the light is focused in front of the mirror instead of behind. So the trick is to direct the light to where it can be viewed – either back through a hole in the main mirror, or away to the side. A bigger practical challenge was producing mirrors polished to the right shape, ideally parabolic rather than spherical. To make matters worse, any imperfections in the surface and shape of the mirror are far more damaging to the final image than for a lens. It was this problem that defeated Reeve, a lens specialist more familiar with glass than metal.

At almost the same time as Gregory was trying to make a working reflecting telescope, the young Isaac Newton was carrying out his first ground-breaking investigations into optics. These led him to the conclusion that white light is made from a mixture of different colours, which are separated when it passes through a prism or lens. He argued that this was the cause of the coloured fringes that had plagued lens-based telescopes since the instrument's invention, and it meant that it was impossible to correct the defect, now known as chromatic aberration. In other words, Newton said that a telescope that used lenses could never give a perfect image. Although this turned out to be incorrect, its pedigree as coming from one of the great thinkers of the age meant that many believed the future of the telescope must lie in another direction.

Pursuing this line of thought himself, Newton turned his attention to creating a telescope with mirrors, since reflection does not split light into its component colours. Rather than seeking the help of London's craft workers, Newton began making it himself, applying his own practical and chemical knowledge. The key, he realized, was the grinding and polishing of a curved mirror (or speculum) that would bring the light to a sharp focus. What he needed was a metal alloy that would be hard and durable, and which would polish well and have a suitable colour. Many trials later, he settled on a mixture of copper and tin, which he used for his first reflecting telescope, completed in 1668. Encouraged, he made two more, and by 1671–72 had improved the alloy by adding a little arsenic to give the mirror a more stable finish. One of the newer examples was examined in the King's presence by Robert Hooke and Christopher Wren. Newton was immediately recommended for membership of the Royal Society in recognition of his achievement.

Despite Newton's initial success, it was not until the following century that the reflecting telescope became a commercial reality. It was the mirrors that proved difficult because speculum metal was so much harder to grind and polish than glass and tarnished quickly. Then in 1721 John Hadley showed his fellow members of the Royal Society a reflecting telescope to Newton's design that he had made with his brothers George and Henry. It could magnify over one hundred times and gave a clear image, because they had successfully ground a mirror with a parabolic curve, the best shape for focusing light. Newton had only managed to produce spherical mirrors, so this was a step further.

The Hadleys' telescope was then put through its paces in a test of optical strength. Its competition was a refracting telescope with the best lens in England. The lens had actually been made nearly forty years earlier by Christiaan and Constantijn Huygens and given to the Royal Society. James Pound then borrowed it in 1717 and mounted it on a maypole in Wanstead Park in Essex, creating an aerial telescope with a focal length of 123 ft (37.5 m). As a well-known and skilled observer, Pound was asked to compare the two telescopes, which he did with his nephew, James Bradley (a future Astronomer Royal). They were impressed. Although only 5 ft (1.5 m) long, the reflecting telescope performed just as well as the refractor for observations of Jupiter and Saturn and was much easier to use. This was the future, Pound concluded.

Although Hadley showed that Newton's system could produce a high-class instrument, the first commercial development of the reflecting telescope was of James Gregory's design. This was due to the work of another Scot, James Short. Intending to go into the church, Short was encouraged to pursue his scientific interests by one of his professors at Edinburgh University and to begin making reflecting telescopes. It was something he turned out to be remarkably good at. Elected to the Royal Society in 1737, Short moved to London the following year and began to sell his telescopes. Unlike other makers and retailers who sold a wide range of scientific instruments, Short proudly styled himself an 'Optician solely for Reflecting Telescopes'. Buying telescope bodies and accessories from other makers, Short fitted them with mirrors that he made himself using a process he kept secret, even ordering his tools to be destroyed before his death. It was obviously a process that worked, because

19   Reflecting telescope, by James Short, London, *c.* 1752–56.

the outstanding quality of his mirrors allowed Short to sell his telescopes for two or three times the price of those of his competitors (Fig. 19). And he was never short of buyers, who included European observatories, serious astronomers and wealthy dabblers. In doing so, he became rich from the sale of just one product, leaving an estate valued at over £20,000 on his death (the equivalent of over £2 million today).

Short made the reflecting telescope into a desirable and profitable commodity. A few decades later it rose to fame as an important astronomical tool, thanks to the achievements of William Herschel and his family. Herschel (Fig. 20) is remembered today as the man who discovered the planet Uranus and as a maker of large telescopes. But he never intended to be either an astronomer or a telescope-maker. Originally christened Friedrich Wilhelm in his native Hanover (in modern-day Germany), he trained from an early age as a professional musician, just like his father and many of his family.

20 William Herschel (1738–1822), engraving by James Godby, 1814, from a painting by Frederick Rehburg.

21 Caroline Herschel (1750–1848), from an oil painting of 1829 by M.F. Tielemann.

William arrived in England towards the end of 1757, having fled from the fighting of the Seven Years War. He hoped to make a living in the new land, where there were lucrative opportunities for trained musicians, and devoted his first years to courting potential clients and giving concerts and lessons. The hard work gained him the position of organist at the newly built and exclusive Octagon Chapel in fashionable Bath, a spa town in the southwest of England. By 1770 he was financially secure and could bring more of his family from Hanover.

First to join William was his younger brother, Alexander, a musician and talented metalworker who built clocks as a hobby. Alexander was followed in 1772 by their sister, Caroline (Fig. 21), who came to train as a singer. She also came with a range of domestic skills, which had been the focus of what education she had been permitted in Hanover. Canny William soon made use of this, since it allowed him to give up the time-consuming business of running a house.

As his musical career gained momentum, William's boyhood interest in astronomy was also rekindled. He began observing shortly after his arrival in Bath in 1766 and soon was buying books on astronomy and related subjects, reading them late into the night. These included one of the most influential eighteenth-century works on optical matters, Robert Smith's *A Compleat System of Opticks* (1738). As he read more, Herschel began making his own instruments. The first was a large refracting telescope, but he was unable to steady it and so turned to John Hadley's step-by-step guide to building a reflecting telescope that appeared in Smith's treatise. Then in 1773 he bought the necessary tools and materials from John Michel, a local Quaker who had given up trying to make telescope mirrors, but who was happy to teach William all he knew of the art. Armed with these tools and Smith's book, he set to work in earnest.

As he began to make reflectors, Herschel concentrated on the mirrors, since the quality of the mirror was the key to a good telescope. Unlike the manual labour of cabinet-making, making a mirror was considered an appropriate activity for a gentleman with an interest in science, as it had been for Isaac Newton. Coupled with his interest in astronomy, this helped to establish William Herschel as a gentleman and man of learning, and would ensure that he was recognized as the 'maker' of the telescopes his family produced in later years.

William's first reflecting telescopes had focal lengths up to about 7–10 ft (2–3 m), with wooden tubes and stands (Fig. 22). As William concentrated on shaping and polishing the mirrors, his brother Alexander made the metalwork pieces, such as the eye-pieces, and a local cabinet-maker produced the tubes and frames. Caroline in turn helped by taking on more of William's domestic and musical duties, including training the local choir. She watched as William's hobby took over the house:

22 A 7-ft (2.1-m) reflecting telescope by the Herschels, c. 1770–90.

> To my sorrow I saw almost every room turned into a workshop. A cabinet maker making a tube and stands of all descriptions in a handsome furnished drawing room. Alex putting up a huge turning machine ... in a bedroom ...

Caroline played another key role by mediating between her brothers, explaining William's instructions to Alexander and Alexander's concerns about what was possible back to William. The brothers themselves seemed too engrossed in their work to understand one another properly. She also became directly involved in telescope-making, later recalling how she helped with the mirror moulds, which were made of 'loam prepared from horse dung of which an immense quantity was to be pounded in a mortar and sifted through a fine sief; it was an endless piece of work and served me for many hours exercise'.

Other tasks were not as unsavoury, but were dangerous. When a mould broke during a mirror-casting, Caroline could only watch in horror as 'my Brothers, and the caster and his men were obliged to run out at opposite doors, for the stone flooring . . . flew about in all directions as high as the ceiling'.

William was taking these risks because size mattered for the type of astronomy he wanted to do. 'The sole end of the work', he said, 'would be to produce an Instrument, that should answer the end of inspecting the Heavens, in order more fully to ascertain their construction.' In thinking about the nature of the stars, William was engaging in a different astronomical programme from other astronomers of the time. The observatories of Europe, including the Royal Observatory in Greenwich, were still working to create better star maps and more accurate predictions of the movements of the heavenly bodies. Their aim was to make navigation more reliable and to help confirm the model of the universe that Newton had set out almost a century earlier. What they carried out was positional astronomy – measuring more and more exactly where the celestial bodies were and how they moved – and they were generally satisfied with refracting telescopes for this purpose. William's quest needed an instrument that would allow him to focus on the nature of the stars themselves. This had to be a reflecting telescope with a large and well made mirror that could gather more light than a lens.

With William seeking bigger and more perfect mirrors, the Herschels were soon building reflecting telescopes of an unprecedented scale. By July 1776, they had constructed a telescope with a 20-ft (6.1-m) focal length. This had to be hung from a mast much like the long refracting telescopes of the previous century, and needed a ladder for viewing. A few years later, William oversaw the construction of an even larger instrument (also with a focal length of 20 ft), and began observing with it in October 1783. This had a sturdy scaffolding framework from which the telescope was suspended, with a platform for the observer built into the frame and a system of ropes and pulleys for moving the tube to the required position (see Fig. 39). Alexander was most impressed with its 'magnificent & stupendous' mechanical mountings. Future telescopes would adopt the same pattern.

William developed new observing methods too. Most astronomers were plotting the positions of fixed stars by measuring and timing them as they crossed their telescope's line of sight. William took to 'sweeping' the sky section by section, looking closely at each point of light to see if it was a single star or something else. He then catalogued them as single stars, double stars, clusters or nebulae (clouds of light-emitting material). This was never an easy task. Observing with his 'small' 20-ft telescope meant standing exposed to the elements at the top of the ladder; on a cold January night in 1783 William found his ink frozen and had to hold the bottle to warm it up.

But the nights of discomfort paid off. During a sweep in March 1781, Herschel recorded 'a curious either Nebulous Star or perhaps a Comet'. Calculations of its orbit showed that the new object was in fact a planet, the first to be discovered since antiquity. In a move similar to that of Galileo and his proposed naming of the Medicean Stars, Herschel wished to call it Georgium Sidus, after King George III, but the name finally chosen was Uranus, after the Greek sky god.

The discovery of the newest planet brought instant fame in magazines and journals, all interested in knowing more about the man behind it, although many writers had difficulties with William's Hanoverian surname, which became Mersthel or Horochelle. As an international celebrity, any subsequent discovery was of interest and might even lead to flights of fancy. After Herschel made a new calculation of the Earth–Sun distance in 1785, Horace Walpole wrote to his close friend Lady Ossory that,

> I revere a telescope's eyes that can see so far! What a pity that no Newton should have thought of improving instruments for hearing too! If a glass can penetrate 1,710,000,000 miles beyond the sun, how easy to form a trumpet like Sir Joshua Reynolds's by which one might overhear what is said in Mercury and Venus that are within a stone's throw of us.

> Well, such things will be discovered – but, alas! We live in such an early age of the world, that nothing is brought to perfection! I don't doubt but that there will be invented spying-glasses for seeing the thoughts – and then a new kind of stucco for concealing them.

Finding the new planet also earned Herschel the respect of astronomers world-wide; he and his instruments were now taken seriously. The following May, he was invited to the Royal Observatory in Greenwich to compare his instruments with those of Nevil Maskelyne, the Astronomer Royal. With them was Alexander Aubert, a wealthy and experienced private astronomer. The results showed the quality of Herschel's instruments. 'We have compared our telescopes together', he boasted to Caroline, 'and mine was found very superior to any of the Royal Observatory.'

This demonstration and William's startling planetary discovery clearly changed Maskelyne's opinion of reflecting telescopes. The Astronomer Royal previously complained of the blurred images they produced, and thought that they should be 'banished from astronomical uses'. Having seen them in action, he was happy to announce that Herschel's reflectors were 'superior to any telescopes made before'.

The discovery of Uranus paid off in other ways, as William's friends worked to find him a royal position. George III had established a private observatory in Kew in 1769 to allow him to view the transit of Venus that year, and there was now an opening since the King's astronomer had recently died. Unfortunately, the post was promised to his son. But William's friends kept up the pressure and he was offered a new post of private astronomer to the King, with an annual pension of £200.

The money allowed William and Caroline to move to Slough in Berkshire, give up music as a profession and devote themselves to astronomy. Caroline found that she was to be trained as William's astronomical assistant and was given a telescope to carry out her own sweeps of the sky. Several months of training later, she began her new work of noting down William's observations and the times at which they were made.

Alexander remained in Bath, since William's earnings were not sufficient to support him as well. Having married and settled there, he found musical work during the Bath season, but discovered that William still desperately needed his metalworking skills. William did try to find a replacement in Slough, but complained that the 'brass man (as we call him) is a stupid fellow'. So Alexander came to stay at the end of the Bath season each year and continued working on William's telescopes.

The official demands on William's time were not heavy. From time to time he had to take his telescopes to nearby Windsor Castle to show members of the King's household the latest astronomical sights, or he might be asked to entertain royal guests sent over to Slough. But the pension it came with was significantly less than the £300 or more he could earn from music. As a supplement, William decided to build telescopes for sale, since people seemed more than willing to pay for an instrument made by the man who had discovered a new planet. This successful move earned the Herschels over £15,000 from the sale of reflecting telescopes and mirrors over the next three decades.

The list of astronomers and observatories they supplied was extensive. Nevil Maskelyne and John Pond, successive Astronomers Royal at Greenwich, and other observatories in Britain and Europe, including Palermo, Göttingen, Madrid and St Petersburg, all purchased their reflectors. A delighted Georg Christoph Lichtenberg reported that Göttingen's new telescope was 'the most perfect work of this art'. Wealthy individuals also bought this latest must-have technology for private use. A more unusual commission was for Walmer Castle in Kent, overlooking the Dover Straits. William Pitt, the Prime Minister, had a 7-ft (2.1-m) reflector installed there in 1799 to give warning of any attempted invasion from France.

The distribution of his telescopes around the world had a further benefit for Herschel's work and reputation, since it helped overcome some of the doubts about his observations. Jean Dominique de Cassini, fourth Director of the Paris Observatory, confided to Herschel that 'many fear that your telescope is subject to optical illusions'. Cassini's words echoed some of the reactions to Galileo's first discoveries, and similar worries had continued to plague telescopic astronomy thereafter (and would do so after Herschel's time). Telescopes were still known to have optical defects, even those by the best makers. An internal reflection once led James Short to claim that Venus had a small moon that exhibited exactly the same phases as the planet. By making and selling telescopes good enough to repeat his own observations, Herschel, like Galileo before him, ensured that other astronomers verified and accepted them.

As he continued his studies of stars and nebulae, William began to yearn for greater seeing power, and soon returned to his royal patron for money to build an enormous new telescope with a focal length of 40 ft (12.2 m). This was given readily at first, but as the costs mounted William was forced to return again and again for more money. As the King grew anxious over the escalating costs, William suggested as a final solution that he grant a lifelong pension to Caroline of £50 per annum and a £200 annual operating allowance for the telescope (on top of the £4,000 spent on its construction). The King agreed on the understanding that there would be no further money.

This was a wonderful moment for Caroline. The pension gave her financial independence while safeguarding her position within the family business, both of which had been concerns after William's marriage to Mary Pitt in 1788. In fact, the marriage strengthened Caroline's position. Relinquishing her role as lady of the house, she was free of the time-consuming duties associated with it and could devote more time to sweeping the heavens and cataloguing stars. From then on, her astronomical successes were prolific and included the publication of an index and corrections to John Flamsteed's star catalogue, the *Historia Coelestis*.

Caroline's sky-sweeping also led to new discoveries: many nebulae and eight new comets. It was the comets that earned her the greatest acclaim. Astronomers were keen to study as many as possible, since detailed observations of their movements were needed to help confirm Newton's theories. Caroline's unmatched skill in discovering comets gradually earned the respect of astronomers world-wide, and was eventually recognized through honorary membership of the Royal Astronomical Society. The public were also fascinated. As the writer Fanny Burney commented on the first comet Caroline discovered, 'The comet was very small, and had nothing grand or striking in its appearance; but it is the first lady's comet, and I was very desirous to see it.'

Meanwhile, William's 40-ft telescope was attracting attention and visitors even before it was finished. Many came from Windsor as part of the entertainment of royal guests. On one occasion, the King led the Archbishop of Canterbury through the unfinished tube, remarking, 'Come, my Lord Bishop, I will show you the way to Heaven!' Fanny Burney was equally impressed, recalling in her diary that she

> took a walk which will sound to you rather strange: it was through his telescope! and it held me quite upright, and without the least inconvenience; so would it have done had I been dressed in feathers and a bell hoop.

Completed in 1789, the great telescope was an impressive landmark (Fig. 23). Some compared it to the Colossus of Rhodes or the Porcelain Tower of Nankin. The American author, Oliver Wendell Holmes described,

23 William Herschel's 40-ft telescope, illustrated in 1819 in the *Encyclopedia Londinensis*. The large supporting framework housed two huts, one for Caroline, the second for a workman. The observer sat in a moving chair and communicated with the huts through a speaking-tube.

OPTICS. Plate IX.

Herschel's Grand Forty-feet Reflecting Telescope.

> A mighty bewilderment of slanted masts, spars and ladders and ropes, from the midst of which a vast tube, looking as if it might be a piece of ordnance such as the revolted angels battered the walls of Heaven with... lifted its mighty muzzle defiantly towards the sky.

For decades afterwards, the Herschels' home became a stopping point on tours of the area and the telescope was marked on the 1830 Ordnance Survey map.

Despite the public interest, the 40-ft telescope was not used for much serious astronomy, although William did try to justify its expense. In a paper published not long after its completion, he claimed to have discovered two moons of Saturn with it, but it seems that he first saw them with his 20-ft telescope. In reality, the 40-ft was cumbersome and difficult to use, taking so long to set up that it ate into each night's observing time. The telescope also needed two operators in addition to William observing and Caroline recording, as well as ongoing maintenance from a large team of workers. Despite the effort and expense that had gone into building the 40-ft, the Herschels preferred their smaller instruments.

Yet the telescope had a significant legacy. Under Herschel's guidance, instruments for investigating space became huge machines, barely recognizable as telescopes, but readily capturing the popular imagination. The 40-ft ignited a desire for ever larger telescopes to gather more and more light. It also provided British astronomy with an iconic image and the emblem of the Royal Astronomical Society.

The story had a more humble end for the telescope itself. William and Caroline used the 40-ft less and less. William's son John didn't use it at all and the telescope remained outside at Slough, unused and unloved. After years of neglect the frame began to rot, so John Herschel decided to have it dismantled. He gathered his family in the telescope's tube on New Year's Day 1840 and together they sang a specially written requiem:

> In the old Telescope's tube we sit,
> And the shades of the past around us flit;
> His requiem sung we with shout and din,
> While the old year goes out and the new comes in.
> Merrily, merrily, let us all sing,
> And make the old telescope rattle and ring!

The great tube was then filled with William's telescope-making apparatus and sealed up. So it remained for thirty years, until a falling tree destroyed all but a quarter. That remaining piece sits in the grounds of the Royal Observatory, Greenwich.

It was an ignominious end for an ambitiously conceived telescope. But the 40-ft was only one part of the Herschels' legacy. Having made their name with important discoveries with reflecting telescopes, William and Caroline helped to create a new direction for astronomical research and to establish the use of large reflecting telescopes in that field. The repercussions for future generations would be significant.

Chapter Five

PERFECTING THE REFRACTOR 1700–1800

> Ah me! my brain begins to swim!—
> The world is growing rather dim;
> The steeples and the trees—
> My wife is getting very small!
> I cannot see my babe at all!—
> The Dollond, if you please!

Thomas Hood, *Ode to Mr Graham, the Aeronaut* (1825)

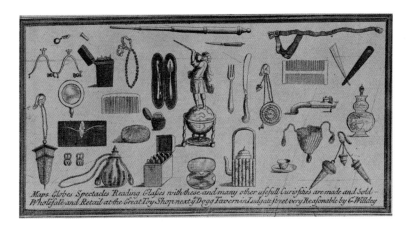

24 The 'toys' sold by George Willdey, a London retailer of the early eighteenth century, from Willdey's *Atlas* (London, *c.* 1714). 'Toys' were small items for adults, such as fans, snuff boxes, writing equipment and telescopes, aimed at an expanding market.

At the beginning of the eighteenth century the refracting telescope was developing along two distinct paths. On the one hand, telescopes for astronomy were becoming longer and often more cumbersome in a quest for greater seeing power. On the other, a range of compact forms was available for use on land and at sea, whether as tools for the professional, or fashionable accessories for the wealthy (Figs 24 and 25). Both forms were to change dramatically over the course of the century. And just as the Herschel family helped bring the reflector to new heights, so a second family with Continental origins helped revolutionize the refractor.

By the middle of the century, London's scientific instrument makers were beginning to monopolize trade throughout Europe and in America. Leading manufacturers like George Graham and John Bird ran firms making the best bespoke instruments for observatories throughout Europe, while others produced everyday items such as spectacles and spyglasses. It was one of these optical firms, set up by the Dollond family, that was central to the development of the refracting telescope.

25 Optical compendium, *c.* 1700. The instruments include a simple telescope, a flea-glass and a lens that creates multiple images.

The story starts with John Dollond (Fig. 26), who was born in Spitalfields in east London. He was the third child and eldest son of a Huguenot (French Protestant) weaver who fled from France with his wife following the 1685 revocation of the Edict of Nantes, which previously protected Huguenots in a predominantly Catholic country. Brought up as a silk weaver like his father, John worked to support the family from an early age, but as a teenager he began to indulge a personal interest in mathematics, particularly astronomy and optics. When he became a father, he passed this enthusiasm on to his eldest son, Peter.

Peter Dollond also learned silk weaving, but never took it up as a trade. Building on the passion inherited from his father, and perhaps a little of the ambition John instilled in him, he set up a

26 John Dollond (1707–61), engraving by K. Mackenzie from a portrait by Benjamin Wilson. Dollond holds a set of prisms used to demonstrate his method for eliminating chromatic aberration.

small optical business. This was successful enough for his father to abandon weaving in 1752 and join Peter in a partnership. A few years later the Dollonds opened their new shop, which proudly advertised itself with the sign of the 'Golden Spectacles and Sea Quadrant', near Exeter Exchange on the Strand, Westminster.

Most instrument making took place in separate establishments dotted around the city, with makers of the different components supplying those who assembled and sold the final products. These sub-contractors included makers of the main operational parts, such as the lenses, and a host of other trades: case-makers, cabinet-makers, and turners in wood, ivory and silver, to name but a few. Most of them worked in small, cramped spaces, either in their own home, or in a small adjoining workshop in a yard in one of the alleys tucked away from London's fashionable streets.

While those at the bottom of this chain were considered mere manual workers, those higher up might pursue the theoretical aspects of their profession. As Robert Campbell wrote in 1747, the scientific instrument maker 'ought to have a Mathematically tuned Head and be acquainted with the Theory and Principles upon which his several Instruments are constructed, as well as the practical use of them'. The director of the Paris Observatory despaired that this was not the case in France, where 'the inferiority of our makers compared with the English comes from their profound theoretical ignorance; the Ramsdens, the Dollonds are geometers and physicists, our best makers are but workmen'. So while it remained a craft and a trade, instrument making had natural associations with science and those interested in it, including the lofty members of the Royal Society. Instrument makers could justifiably claim a place in this genteel world. By the 1750s, John Dollond had successfully done so, as a member of the Spitalfields Mathematical Society and a regular correspondent with members of the Royal Society and scientists throughout Europe.

Pursuing his interest in optics, John began to look at how to improve refracting telescopes. The big problem was chromatic aberration. The great Isaac Newton said this could never be solved, but several thinkers were beginning to question whether he was correct. The Swiss mathematician Leonhard Euler argued that Newton was wrong because nature had produced a lens with no such defect, the eye, reviving an argument previously put forward by David Gregory. Then in 1754 the Swedish astronomer Samuel Klingenstierna published a paper that disproved the conclusions Newton had drawn from his experiments. Dollond initially defended Newton, but a closer reading of Klingenstierna's arguments persuaded him that a practical solution to chromatic aberration could be found.

Dollond's further research bore fruit in a patent issued on 19 April 1758. This was for a new lens made from two types of glass (crown and flint), which Dollond claimed was free of both chromatic and spherical aberration. Chromatic aberration was eliminated because the different types of glass refracted the light differently and could be combined in such a way that the light was not split into its component colours. Spherical aberration was eliminated, the patent said, by ensuring that the shapes of the two parts of the lens worked together to ensure that the light was brought to a focus at a single point.

Ten days later the *Daily Advertiser* publicized a 'new-invented Refracting Telescope, by which Objects are seen much clearer and distincter than by any before made'. Dollond proudly wrote to his friend James Short that he could make refracting telescopes that, 'in the opinion of the best and undeniable judges . . . far exceed anything that has hitherto been produced'. Later the same year he also presented a paper to the Royal Society, describing his solution to the problem of chromatic aberration, although not the details of making a lens: the paper discussed his ideas only in terms of glass prisms. The Society must have been impressed. John was awarded its highest honour, the Copley Medal, and made a Fellow three years later.

By this time Dollond had an agreement with another instrument maker, Francis Watkins, to sell telescopes with the new lens (Fig. 27). By largely solving a problem that had plagued refracting telescopes for 150 years, the two-part lens meant that reliable telescopes could now be made in a range of convenient sizes. The potential for profit must have seemed huge, but the secret was soon discovered and similar telescopes by other London opticians began to appear. Within a year, a retailer named Benjamin Martin was calling the new telescopes 'achromatic'. The name stuck.

John Dollond remained fiercely proud of his discovery, and telescopes with achromatic lenses were a commercial hit. But after John's death in 1761, Peter Dollond made an aggressive move to claim the full financial benefit of the invention by enforcing his father's patent. Once he did so, he did it with a vengeance through

27   Spyglass, by Watkins, London, *c.* 1760. Made by the firm set up by Francis Watkins, who was in partnership with John Dollond by this time.

a series of court cases beginning in 1763. Ironically, it was Francis Watkins whom he first sued. Over the next few years Dollond successfully fought twelve lawsuits against seven opticians, driving a number out of business and gaining control of the manufacture and retail of achromatic lenses until the patent expired in 1772.

During these court cases and the resulting counter-claims by a group of London opticians, some doubt emerged as to John Dollond's true role in the invention of the achromatic lens. The London opticians claimed that John had merely developed the ideas of Chester Moor Hall, a lawyer from Essex. Hall successfully developed lenses that reduced chromatic aberration in the 1730s, with Dollond hearing about them from George Bass, a lens-grinder. According to this account, Hall commissioned two London opticians, Edward Scarlett and James Mann, each to make one part of the lens. Both unknowingly sub-contracted the work to Bass, who realized that the two must fit together.

This story may never be completely untangled, but there are striking similarities to Galileo's role in the telescope's early history. Like Galileo, the Dollonds took a promising idea by someone else and made it a practical reality, in their case a highly commercial one.

It was precisely this point that Lord Camden, Chief Justice of the Court of Common Pleas, reiterated in Peter Dollond's case against James Champneys in 1766. The judge ruled that, 'It is not the person who locks his invention in his scrutoire who ought to profit by a patent for such invention, but he who brings it forth for the benefit of the public.' The Dollonds had certainly done that.

Whether or not John was the first inventor, he and Peter successfully pioneered the achromatic telescope, building a sizeable business in the process. It is perhaps no surprise that the lawsuits of the 1760s created bad blood between Peter Dollond and the other London opticians, but by the end of the decade things had been smoothed over and his family was becoming an important part of the scientific instrument making community. This was important for the firm's growth, since London's instrument makers formed an extensive network fostered by a formal system of apprenticeship. Under this system, boys (and some girls) trained with master craftsmen to learn their skills. This fostered long-term associations, with some apprentices marrying into their master's family and eventually taking over the business.

As a member of the Spectaclemakers' Company, Peter took on apprentices, creating lasting links as each completed their training and went into business for themselves. John had also worked to form ties with his commercial counterparts, arranging the marriage of his two daughters to instrument makers, Sarah to Jesse Ramsden and Susan to William Huggins. Sarah's marriage should have been a coup, since it brought together what were to become two of the biggest names in the London trade. Sadly, it lasted just seven years, although Sarah carried on managing parts of the Ramsden business. Susan and William Huggins's marriage was much more successful. Their son George trained under his uncle Peter and subsequently took over the firm, even changing his surname to Dollond. It was important to keep a Dollond in charge.

Both during the lifetime of the Dollond patent and after 1772, when prices dropped significantly, the firm was enormously successful and became renowned for the quality of their products. This was helped in 1763 by Peter's introduction of a newer form of achromatic lens with three glass elements instead of two. These triplet lenses proved even better at reducing the effects of aberration, helping to cement the firm's reputation. Dollond soon became the only name from whom the discerning customer would buy a telescope. A few weeks before Lord Nelson's death at the Battle of Trafalgar, the artist Benjamin Haydon passed him by in London and noted that the admiral 'had been to Dollonds to buy a night-glass'. Marie de Vichy-Chamrond even complained to Horace Walpole, who unfortunately sent her a telescope by another maker, that, 'I want one of M Dollond's excellent telescopes which enlarge objects eighty times or at least sixty times. The one you have sent me magnifies only ten times.'

As the company expanded in later years, the Dollonds revolutionized telescope design in other ways. They began to use mahogany for the barrels, since it was stronger, and introduced brass draw tubes, which became standard for telescopes sold everywhere (Fig. 28). Through innovation and successful marketing, the Dollonds sold thousands of telescopes to gentlemen and to professionals alike. As the firm's reputation and sales grew, a 'Dollond' became a synonym for a telescope and passed into the slang of the Royal Navy.

The firm also operated at the top end of the market, supplying bespoke instruments and components to astronomers and observatories almost everywhere. Nevil Maskelyne, fifth Astronomer Royal at Greenwich, was well ahead of the game. He bought his first achromatic telescope not long after starting in 1765; it was one of the first ever made with one of Dollond's triplet lenses. Among his later purchases was a telescope with the largest achromatic lens Peter Dollond had ever made, 5-in. (12.7-cm) in diameter (see Fig. 29). The advantages must have been obvious, since Dollond soon became the Observatory's main supplier of lenses, many used to upgrade the older instruments there. Observatories overseas

28 Portable telescope, by Dollond, London, *c.* 1810. This compact model shows two features introduced by the Dollonds: the wooden barrel and sliding brass draw tubes.

29 Transit telescope by Troughton, London, made for the Royal Observatory, Greenwich in 1816. The telescope incorporated Peter Dollond's largest achromatic lens, 5 in. (12.7 cm) in diameter, made for a different instrument twenty years earlier.

were also just as keen on Dollond's new refractors, which rendered obsolete astronomical telescopes without the new lenses.

It is difficult to overplay the importance of the achromatic lens. Commercially, it made the fortunes of a firm whose successful legacy remains in the British high-street opticians, Dollond & Aitchison. But its long-term significance was greater. In the century after London opticians finished fighting over the rights to produce it, the achromatic lens underpinned a huge diversification in telescopes and telescopically aided instruments as Europeans and their instruments began to straddle the globe.

Chapter Six

GLOBAL DOMINATION 1720–1900

I searched him, and found the telescope in his breeches.

John Hutt at the trial of Thomas Jones and James Hunter for stealing a telescope worth 16 shillings in 1818. Jones, aged 15, was transported for seven years; Hunter, 13, was put in prison for two months.

30   *More Ships or Good News from Copenhagen!!*, published by Thomas Tegg, 1807. Satirical prints depicting King George III made fun of his poor eyesight, for which he carried a spyglass, a common fashion accessory by this time. The spyglass, the prints suggested, further distanced the King from the world around him.

The work of the Dollonds and other makers introduced many features that would characterize telescopes for the next century and more. The achromatic lens became standard, as did brass draw tubes and wood-covered barrels. But within this broad standardization, there was room for diversification and specialization as the telescope spread geographically and into different professional and personal areas.

The introduction of the achromatic lens allowed makers to produce a bewildering range of optical instruments in almost any size. Even at the beginning of the eighteenth century, telescopes were being sold as items of intellectual amusement to the well-off (see Fig. 25). This proved to be an expanding market, reflecting the growth of a consumer culture that flaunted its wealth through fash-

31 *An Irish Pilot or Steering by Chance*, published by Thomas Tegg, c. 1812.

ionable accessories, including spyglasses and other visual aids (Fig. 30). In the decades after 1758, an even wider range of telescopes was developed, for men and women of fashion, soldiers and sailors, and those who wished to study the skies. The new telescopes were also added as sighting aids on other instruments, including the navigational and surveying equipment developed in the late eighteenth century.

A good guide to the range available is William Kitchiner, an avid telescope collector whose eclectic interests included food and music; among other things, he composed patriotic and nautical songs. In *The Economy of the Eyes* (1824–25) Kitchiner pays some attention to opera glasses, simple Galilean spyglasses (binocular versions appeared later in the century). He recommends Dollond's Bell Opera Glass, which is as 'an entertaining companion at a Play-house, and a very pleasant Prospect-glass', but warns that other examples are nowhere near as useful. These 'toys' may be pretty, he cautions his readers, but are 'made to be looked at rather than to be looked through'.

Kitchiner also reveals the fascinating selection of accessories available to the well-heeled man or woman about town. These include different types of perspective and day telescopes that fold up to a convenient size. The 'Invisible Opera Glass' is ideal for travellers wishing to examine architectural details. For the more secretive, the 'circumspector' and 'diagonal opera glasses' allow one to examine other people without being noticed. In fashionable circles, after all, not being seen to look was just as important as watching others. Fans concealing small telescopes catered for this same need.

Kitchiner also discusses hand-held telescopes for the professional. Mariners had been important customers since the seventeenth century (Fig. 31). Kitchiner reiterates the point that every officer must have a good telescope, for it is 'the means of safety either from the dangerous effects of a storm, or an enemy's ship in sight'. And there were many available. The Deck Glass or Look-out was small enough to be carried aloft. The Night or Day telescope was said to be useful in poor visibility, or perhaps better the Nightglass with its large objective lens to gather more light.

In the later nineteenth century these specialized marine telescopes became more standardized. By the end of the century the Deck Glass had become the Officer of the Watch telescope (Fig. 32). This was made to an Admiralty pattern and tested and certified at the Royal Observatory or at Kew Observatory.

On land, army officers were another growing market. Canny businessmen like Dollond and others soon began to produce compact, yet powerful and sturdy, telescopes for their needs (see Fig. 28). In 1778, the New York dealer James Rivington recommended these portable spyglasses as the '*Sine qua non* in RECONNOITERING the FOE, first invented by Mr. DOLLAND [*sic*]'.

32   Officer of the Watch telescope given to L. Donaldson in 1890.

Numerically, these telescopes for civilian and military use dominated production. For retailers like the Dollonds, they formed the core of an expanding and successful business. In the end, they proved so profitable that some retailers capitalized on the success of the most famous makers by selling telescopes bearing suspiciously similar names. Towards the end of the nineteenth century this reached the point where Dollond & Co. felt compelled to warn buyers that telescopes marked 'Dolland' were frauds.

While the manufacture and retail of hand-held telescopes had a high turnover and proved to be a lucrative trade, astronomers were still a key market for state-of-the-art, bespoke telescopes. These were high-value commissions that could make or break the reputation of optical firms, so their custom was worth carefully pursuing.

By the end of the eighteenth century, astronomical telescopes were developing in two directions, reflecting the needs of two distinct research programmes. On the one hand, astronomers in state-funded observatories were working to improve star maps.

Refracting telescopes made with the new achromatic lenses suited their needs well. On the other, the excitement of William Herschel's discoveries encouraged privately funded astronomers to look more closely at the stars themselves. It was the greater light-gathering abilities of reflecting telescopes that proved more seductive to these observers, a number of whom could afford to spend large sums of money and considerable time and effort to build telescopes and observatories to rival those funded by the state. Nevil Maskelyne neatly summarized the difference between the two types of astronomer when he said that, 'Mr. Herschel looks at the appearances of bodies more than the times': accurate timings underpinned the precision mapping of the stars that he and his peers in the national observatories were undertaking. The contrast becomes even clearer by comparing the work of the Royal Observatory at Greenwich with that of one private astronomer.

Until the mid-nineteenth century, the Royal Observatory's work was almost exclusively positional astronomy in support of the navigational purposes set out at the time of its foundation. In order to maintain and improve the astronomical tables mariners needed to safely cross the world's oceans, its precision work relied on the highest quality instruments. Each successive Astronomer Royal was therefore keen to commission new telescopes and observing instruments from the best firms. One maker observed that, 'a new instrument was at all times a better cordial . . . [for John Pond, sixth Astronomer Royal] than any which the doctor could supply'. This constant updating of instruments also established the Royal Observatory as a leading and influential player in the development and use of astronomical telescopes elsewhere.

This pattern was established early on. In September 1725, Edmond Halley, second Astronomer Royal, was delighted as he took delivery of a large mural quadrant by George Graham, one of London's leading makers. The quadrant combined a telescope with an accurate degree scale and became Halley's principal observing instrument. Its quality ensured that it was in regular use at

Greenwich until 1812, while the demonstration of its accuracy through Halley's work established it as the prototype for instruments built for other European observatories. When Halley's successor, James Bradley, ordered a second mural quadrant, it too was based on Graham's design (Fig. 33, right) and built by another of the capital's best makers, John Bird.

Bradley arrived at Greenwich with an outstanding reputation in positional astronomy, having discovered the aberration of light and the nutation (irregular motion) of the Earth's axis. The discovery of aberration, the apparent change in the position of observed objects caused by the Earth's motion, was particularly important as the first observational evidence that the Earth orbited the Sun. He made these discoveries with another telescopic instrument, the zenith sector (Fig. 33, left). This was also one of George Graham's creations, and Bradley openly acknowledged the debt he owed to the London maker. Graham's 'great Skill and Judgement in Mechanicks, join'd with a complete and practical Knowledge of the Uses of Astronomical Instruments', he wrote, lay behind the celebrated discoveries. The famous zenith sector came to Greenwich with Bradley and before long was copied by observatories everywhere.

The pattern repeated itself again and again, although Greenwich did not always lead. While Halley installed England's first transit telescope in the Royal Observatory, the instrument was originally designed by a Danish astronomer, Ole Rømer. It too became standard in observatories everywhere. Likewise, the new must-have astronomical instrument of the late eighteenth century was the astronomical circle, a complete divided circle with a telescopic sighting tube. The model example was made by Jesse Ramsden's London firm for Palermo Observatory in 1789. Within three years Nevil Maskelyne was asking for permission to buy one for Greenwich, without success. He tried again in 1806, complaining that 'most of the observatories of Europe are now furnished with divided circles'. This wasn't just instrument-envy but a question of accuracy, since they were better than the mural quadrants he

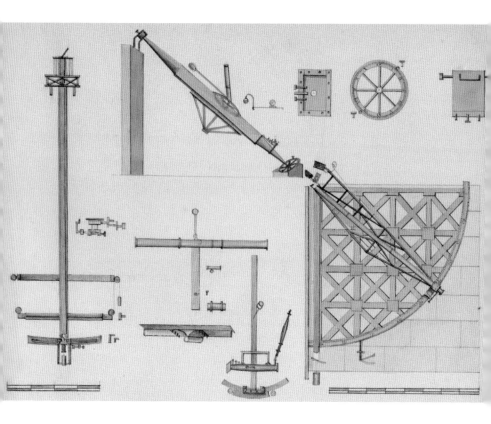

33 Some of the instruments at the Royal Observatory, Greenwich, drawn by John Charnock of Blackheath, *c.* 1785. On the right is the large mural quadrant commissioned by James Bradley. On the left is the zenith sector with which Bradley discovered the aberration of light and nutation of the Earth's axis. The other images are of an equatorial sector and its accessories.

was still using. Maskelyne was finally allowed to order one the following year and it was installed in 1812, by which time John Pond was in charge. Unfortunately, it proved too unsteady for accurate transit observations, so Pond ordered a new transit telescope instead (Fig. 29).

Jesse Ramsden's commission for Palermo was typical of the London trade's near monopoly of the European market in the later eighteenth century, but instrument makers in Germany and France began to challenge and beat London makers in securing prestige commissions in the century that followed. Their success stemmed from improved methods in the manufacture of optical glass and from the work of two pioneers, both with unusual histories.

The first was Pierre Louis Guinand, a Swiss watch- and bell-maker who spent many years experimenting on ways of improving flint glass. His eventual success aroused the interest of many European glass-makers, including Joseph von Utzschneider, who persuaded Guinand to come and work for his recently established Optisches Institut (Optical Institute) near Munich. To guard the secrets of his new techniques, Guinand's contract even stipulated that only he and his wife could operate the glass-making furnaces.

The second figure was Joseph Fraunhofer, who by the age of twelve was an orphan apprenticed to a decorative glass- and mirror-maker in Munich. A second apparent stroke of misfortune changed his life again, when two years later he was caught in the collapse of the house he was living in. Dragged unconscious from the ruins, he came to the attention of the Elector of Bavaria, who instructed the ubiquitous Utzschneider to ensure the boy's welfare. The young Joseph impressed Utzschneider, who encouraged his growing interest in optics and offered him a junior post in his company in 1806. To Guinand's horror, he was instructed to teach the young man all he knew about glass-making. This proved to be a shrewd move on Utzschenider's part, since Fraunhofer guided the firm to enormous success. It was disastrous for relations with Guinand, who returned to Switzerland in 1814 and started a family business making lenses.

Utzschneider's firm started modestly, producing marine telescopes that copied English designs. But as the technical expertise of Fraunhofer and his colleagues grew, the range and quality of the firm's instruments rapidly improved. Within ten years it had a reputation as a maker of the highest quality. This was helped by a number of prestige projects. One was a large refracting telescope for Dorpat Observatory in Russia (now in Estonia), commissioned by the astronomer Friedrich Georg Wilhelm Struve. Completed at the end of 1824, Struve was so impressed that he wrote to Fraunhofer that he was 'undetermined which to admire most, the beauty and elegance of the workmanship in its minute parts, the soundness of its construction, the ingenious mechanism for moving it, or the incomparable optical power'.

A second prestigious commission, for Königsberg Observatory, further enhanced Fraunhofer's international reputation after Friedrich Bessel used the telescope to detect stellar parallax and measure the distance of a star for the first time in 1838. Although observation of stellar parallax was no longer really needed to prove that the Earth orbited around the Sun, the measurement of stellar distances was essential for calculating the size of the universe. The success was a fitting testament to the quality of the telescope and the observer, Bessel having succeeded where Tycho, Flamsteed, Bradley and other famous astronomers had failed for want of sufficiently accurate instruments.

The successes of Fraunhofer and others soon meant that German firms were leading the way in the production of high-quality astronomical telescopes, with even the Royal Observatory at Greenwich commissioning lenses and telescopes from them. The growing threat to British manufacturing led to desperate attempts to obtain the glass-making secrets in the 1820s: Fraunhofer was offered £25,000, but stuck to his contractual obligation to secrecy. The British turned to Guinand's family, once more without success. They then tried the French opticians to whom Henri Guinand had disclosed his father's secrets. They failed again. The result was that British optical glass-makers could not match their German and French rivals until the 1850s, by which time they had finally learned the new methods.

These fluctuations in fortune and reputation are well illustrated in the new instruments commissioned for Greenwich by George Airy, seventh Astronomer Royal from 1835. Although a man of only medium stature, Airy had great powers of endurance and oversaw a forty-six-year period when the work of the Observatory expanded into new areas of astronomy such as spectroscopy. Yet he always maintained that Greenwich's 'staple and standard' work was the production of accurate astronomical tables and data. Like his predecessors, Airy sought the best new observing equipment, much of which he personally designed. Sometimes this meant ordering components from Germany, including an objective lens by Merz for the 'Great Equatorial' telescope installed in 1859. Other instruments were supplied and built entirely by English firms as they re-established their reputations. A large altazimuth telescope installed in 1847 to observe the Moon's daily positions used optical components by the English firm of Troughton & Simms.

The most famous of Airy's new instruments was a large transit circle (Fig. 34). Having bought a high-quality lens from Simms in 1848, Airy designed an impressive instrument to house it, employing Ransomes & May of Ipswich for the sturdy cast-iron structure and Troughton & Simms for the remaining optical and instrumental parts. At the heart of the new instrument was a telescope that rotated in the plane of the meridian and which could make extremely accurate observations. It was used to make more than 600,000 observations between 1851 and 1954. The Greenwich Meridian was also defined by the telescope, which was used to check Greenwich Mean Time every day. Thirty years later, the 1884 International Meridian Conference in Washington DC decided that the world's Prime Meridian would be defined as the meridian passing through the centre of the telescope. It is the meridian that millions of visitors come to see in Greenwich.

Away from the professional observatories, many wealthy individuals were continuing the astronomical programme begun by William Herschel, putting their money into reflecting telescopes

34 Airy's transit circle in use at the Royal Observatory, Greenwich, from E. Dunkin, *The Midnight Sky* (London, 1891). The observers noted a star's position by recording its angle and the time as the star transited (crossed) the instrument's meridian. The transits of specific 'clock stars' were also used to find Greenwich Mean Time.

with mirrors of ever greater diameters in order to see further into the stellar regions. The most impressive was the so-called Leviathan of Parsonstown, in many ways the culmination of Herschel's programme. This monster telescope was designed and paid for by

William Parsons, the third Earl of Rosse, who had it erected at his ancestral home, Birr Castle in western Ireland.

Parsons had been interested in astronomy since the 1820s and earned his place in the international astronomical community as a member of the Astronomical Society of London (later the Royal Astronomical Society) and the Royal Society, of which he became president. He also regularly published the results of his work, much of it on new techniques for making telescope mirrors.

Determined to push Herschel's work further, Parsons designed and commissioned several reflecting telescopes from the 1830s, hoping to solve unanswered questions about the nature of nebulae. Most of all, he wished to show that every nebula was a group of stars, something that only telescopes with very high resolving powers could reveal. After five attempts, he successfully cast a mirror with a diameter of 6 ft (1.8 m), which was mounted in a framework built between two stone walls. Parsons's sense of expectation as the telescope was made ready for observing in 1845 shone through in a poem written in praise of his 'Lord of Ether and Light':

> Thou shalt lead us on in triumphs,
> Yet to mortal power unknown;
> Realms, which Angels only visit,
> Shall yield homage at thy throne.

These hopes seemed to be fulfilled almost immediately, with the first discovery of a spiral nebula proving that the telescope's mirror was of the highest quality. Word of the Leviathan spread in newspapers, guidebooks and lectures, all heralding a technological wonder of the new age. Harriet Martineau cited it as one of the engineering marvels achieved by thirty years of peace since the Napoleonic Wars. Prints and other images helped to cement its image as a spectacular symbol of technological progress (Fig. 35), and, like Herschel's great 40-ft telescope, it became a tourist attraction.

35 Linen wall hanging of the 'Leviathan of Parsonstown', produced for the Working Men's Educational Union, London, c. 1860. This was one of a number of hangings used for lectures to the working classes. It shows the Leviathan's 56-ft (17-m) tube, the observing platform for use at low elevations and the gallery above for higher elevations.

Parsons and his telescope inspired other wealthy amateurs. The Liverpool brewer William Lassell visited in 1844 to discuss speculum-grinding while planning a reflector with a 24-in. (61-cm) mirror. It was completed two years later and had a major success when Lassell discovered Triton, the largest moon of Neptune, itself discovered just weeks earlier. Lassell went on to build an even bigger instrument and was regularly in contact with James Nasmyth in nearby Manchester. Nasmyth, a Scottish engineer better known for his invention of the steam hammer, also spent his fortune and spare time making large telescopes. His engineering knowledge led him to develop several innovative designs. He also enjoyed telling the tale of the night he scared a bargeman while carrying a telescope into his garden dressed in his nightshirt. He looked like a ghost with a coffin.

Yet although the Leviathan was an inspiration to others and was used to discover further spiral nebulae, it never fulfilled its early promise. Not least of its problems was the Irish weather, which often prevented observation, while the tarnishing of the mirrors diminished its seeing power. Like Herschel's 40-ft, the Leviathan was difficult to manage, needing at least three assistants to move it.

Above all, the Leviathan of Parsonstown symbolized a very different approach to astronomy from that of the state observatories. But these two programmes began to converge in the second half of the nineteenth century as the science of astrophysics evolved. The spark was the development of two new ways of using telescopes to explore the heavens: photography and spectroscopy.

Photography's potential for recording telescopic observations was recognized soon after the creation of the first permanent image in the 1820s, with photographs of the Moon made by John Draper of New York University in 1840. As improved techniques reduced exposure times, the possibilities for detailed recording and measurement by photographic means became obvious. George Phillips Bond delightedly pointed out in 1857 that, 'The plates once secured, can be laid by for future study by daylight and at leisure. The record is there, with no room for doubt or mistakes as to its fidelity.' From his position of Director of the Harvard College Observatory, Bond became one of the main advocates for astronomical photography, suggesting that photographic images could be used to determine stellar magnitudes and positions accurately. 'The photograph', he boasted, 'is worth three times as much as a single direct measure.'

There was another motive behind Bond's words. Many astronomers were coming to believe that photography could finally remove subjectivity from astronomical observation, which had long remained open to personal interpretation and disagreement. This was typical of a range of moves to replace what was seen as fallible human observation with mechanical substitutes. As one author wrote in the *Edinburgh Review* in 1888, the 'stars should henceforth register themselves' through photography without human interference or the vagaries of aesthetics to confuse matters. In the end, the new techniques could not be perfected without subjective standards, even to resolve the question of whether an image was real or an instrumental distortion. But there was good reason for optimism. Photography could reveal far more about the physical nature of the universe, because photographic film could capture more light and more information than the eye. By fixing images for posterity, it also offered a new way of mapping the heavens, and was the basis of an ambitious project launched in 1887. Combining the work of observatories world-wide, the *Carte du Ciel* was to be a photographic catalogue of all stars to the eleventh magnitude, but proved such a mammoth and complex task that it only resulted in a partial catalogue in the 1960s.

The second new technique was spectroscopy, the study of electromagnetic spectra. The absorption and emission of different wavelengths of light by liquids and gases had been investigated in the first half of the nineteenth century. Its use as a tool for astronomical research was most clearly demonstrated by William Huggins, who had a private observatory in Tulse Hill, London. Following the work of Fraunhofer and Gustav Kirchoff, who both analysed the light from celestial bodies, Huggins and William Miller (a professor at King's College, London) turned a telescope with a spectroscopic attachment to the stars. Their findings were amazing. They showed that the stars are made of the same elements as the Earth, which was a revelation in itself, and that some nebulae were made entirely of gas and were not groups of stars at all. This second finding showed that Parsons's expectations of using his Leviathan to identify the individual stars in every nebula could never be realized.

Further pioneering work in the 1860s included that of Father Angelo Secchi of the Collegio Romano, whose analysis of the spectra of more than 4,000 stars led him to propose a new stellar classification scheme. The work of Huggins, Secchi and others established a new way of using telescopes to understand the physical nature of the universe, now shown to be just like that of Earth. Ironically, spectroscopy also allowed scientists to use telescopes to understand the universe at the smallest level possible, that of the elements.

The new research areas opened up by astronomical photography and spectroscopy created a new discipline: astrophysics. It was this new field of study that encouraged state and university observatories to tackle the questions that privately funded astronomers like Parsons were pursuing. It also encouraged a world-wide mania for building large telescopes, although these were still mainly refracting telescopes. Observatories in Europe and America soon appeared to be in a race to build the biggest and best refractor in the world. Things really heated up in the 1870s, with a new record-breaker built every few years: 26 in. (66 cm) at the US Naval Observatory, Washington (1873); 30 in. (76.2 cm) at Pulkovo, near St Petersburg (1885); 36 in. (91.4 cm) at the Lick Observatory on Mount Hamilton, California (1888). It culminated in a refractor with a 40-in. (1.02-m) lens completed in 1897 for the University of Chicago's Yerkes Observatory in Williams Bay, Wisconsin, USA. It is still the world's largest operating refractor.

These new developments in astronomy mostly took place during George Airy's time at the Royal Observatory in Greenwich. The Observatory also broadened its work to include research into the physical nature of the universe, beginning programmes of solar photography and spectroscopic observation. Nor did the Observatory remain immune to the size race: in 1885 Airy's successor, William Christie, commissioned a refracting telescope with an objective lens of 28 in. (71 cm) in diameter, which was installed in the mounting originally built for a telescope Airy had commissioned thirty years previously (Fig. 36).

36 The 28-in. (71 cm) telescope at the Royal Observatory being operated by Thomas Lewis (at the eyepiece) and William Bowyer, c. 1899. Deployed initially for spectroscopic work, the telescope was used to observe double stars for much of its working life. It is still in working order at Greenwich.

Another story unfolded in tandem with these developments in astronomy. During the late eighteenth and nineteenth centuries, Britain extended its imperial reach through the colonization and exploitation of territories overseas. Telescopes had many roles in this expansion. They helped mariners navigate the oceans safely, as navigational instruments and as sighting aids on position-finding instruments like the marine sextant. They were also important tools for the surveyors, hydrographers and astronomers who crossed the oceans to accurately position and map the new territories. Each needed telescopically enhanced surveying tools like theodolites and surveying levels, or portable astronomical instruments for determining accurate geographical positions by celestial observation.

37 *A View at Anamooka*, engraved by W. Byrne after a drawing by John Webber, published 1784. William Bayly, astronomer on the *Discovery* during Captain Cook's third voyage, observes with an astronomical quadrant on one of the Tongan islands in 1777.

Captain Cook's three famous voyages are good examples. As a matter of course, astronomers were sent on each voyage to determine latitudes and longitudes to map the territories Cook was discovering. Latitude was measured from observations of the Sun. Longitude could be determined by observing eclipses, the Moon's position and other celestial phenomena. The main scientific objective of Cook's first voyage (1768–71) was to observe the transit of Venus of 1769 from the Pacific to help work out the distance between the Earth and the Sun. On each voyage the astronomers were handsomely equipped with instruments including reflecting telescopes (and refractors on the second and third voyages), astronomical quadrants and accurate timekeepers (Fig. 37). Unfortunately, they were all valuable items that could be tempting to others. An astronomical quadrant was stolen during the first voyage before it had even been unpacked for observations of the Venus transit. It was later recovered, somewhat damaged. Again on the third voyage (1776–79) a sextant was stolen from William Bayly's temporary observatory on the Polynesian island of Huahine. This was not a matter to be taken lightly: Cook had the thief's ears cut off, but the man escaped a few days later.

Such misadventures aside, the scientific equipment for Cook's expeditions set the template for others in the decades that followed. As a great age of naval exploration and expansion continued into the next century, an expanding range of telescopic and other instruments was designed specifically for accuracy and ease of transportation as scientific expeditions covered the globe (Fig. 38).

In British territories in Asia, the East India Company also made great efforts to map the regions it controlled, building observatories and undertaking extensive land surveys. At Madras Observatory, founded in 1792, the symbolic value of the building and its instruments was spelt out in a boastful inscription on one of the granite piers supporting the large transit telescope. 'Posterity may be informed a thousand years hence of the period when the mathematical sciences were first planted by liberality in Asia.' The message's rhetoric imperiously overlooked Asia's long-established mathematical traditions, which were flourishing in India a mere thirteen centuries earlier.

Observatories sprang up almost everywhere in Britain's expanding empire through state and individual action. In South Africa, the British Admiralty established an observatory at Cape Town in 1820, with a permanent building completed in 1829. Four years later John Herschel sailed to South Africa, at his own expense, with his

38 Portable altazimuth telescope, by Troughton & Simms, London, 1866. One of the instruments used in New Zealand on a British expedition to observe the transit of Venus of 1874.

39 *Site of the Twenty Feet Reflector at Feldhausen, Cape of Good Hope Sept 1834*, from a drawing by John Herschel, published in 1847. The 'large' 20-ft (6.1 m) reflecting telescope was originally built in the 1780s by William Herschel, John's father. Transported into the African landscape, it presented a clear statement of the role of British and European science in 'civilizing' the new territories.

family and telescopes. Helped by the director of the Cape Observatory, he set up his biggest telescope at Feldhausen, a house they were renting near Cape Town, and used it to map the southern skies before returning to England in 1838 (Fig. 39).

Still further from Britain, Thomas Brisbane, colonial governor of New South Wales, Australia, 1821–25, assuaged his passion for astronomy by establishing Sydney's first permanent observatory at Parramatta. Its expensive equipment from the home country included a high-quality transit telescope with which he hoped to realize his dream of a 'Greenwich of the southern hemisphere'. Brisbane's astronomical interests came to symbolize the man himself, something that became clear when his cousin, James Brisbane, came to Sydney in command of the Royal Navy's China squadron. In the course of the visit, Bungaree (or Bongaree), an aborigine well known for his help during Matthew Flinders's circumnavigation of Australia, boarded HMS *Warspite* and was introduced to James Brisbane. Charles Dickens later recounted what happened next, as Bungaree exclaimed that this was not the real Brisbane (the governor):

> taking a telescope from the hand of the signal-midshipman of the day, [he] looked through it into the heavens, and exclaimed, "Ah!"...The commodore was so struck with King Bungaree's imitation of his own first cousin, that he stood aghast; while the officers, unable any longer to preserve their gravity, indulged in a hearty peel of laughter.
> *All the Year Round*, 21 May 1859

To Bungaree, the act of telescopic observation distinguished the governor above all else. But, like Brisbane's term of office, his observatory proved short-lived; it collapsed after being eaten by termites.

The situation was rather different in those lands that remained separate from the growing British empire, as Lord George Macartney discovered when he arrived at the Summer Palace of the Qianlong Emperor of China in 1793. Macartney's brief as head of an embassy heavily funded by the East India Company and the British government was to create new trading links with China. The embassy brought lavish gifts of the finest British manufactures to impress the Chinese court. These included scientific instruments, for which English makers enjoyed a reputation world-wide and which were known to fascinate the Emperor. Twenty years earlier, the Jesuit Michel Benoist described Qianlong's enthusiasm for a telescope he had been given. The Emperor had two eunuchs carry it around at all times and never tired of its 'surprising effects'. With this in mind, Macartney's rich gifts included refracting telescopes by Dollond, a zenith sector by Ramsden, and other surveying and navigational instruments. There was also a large planetarium with Herschel's newly discovered planet Uranus added. It was complemented by a reflecting telescope with one of

Herschel's mirrors, an example of the instrument with which the new discovery had been made. Macartney had purchased the mirror in Canton for £200 as he added further gifts to those originally packed into his ships. To emphasize its importance, the catalogue of presents explained that, 'The powers of vision are extended by the means of Mr. Herschell's machine, beyond the hopes, or calculations of all former philosophers.'

The embassy failed. The Chinese saw the gifts as tribute from a lesser nation in the western ocean. They were viewed as luxurious novelties like the automata and instruments that were already common in court circles. Macartney's claim that they were the keys to tremendous advances in knowledge and to the mastery of nature fell on deaf ears.

Although the Chinese did not believe it, Europeans no longer questioned the idea that scientific and technological progress exemplified their cultural superiority world-wide. It was a message that lay at the heart of the national and imperial posturing of the nineteenth century. Nowhere was this more evident than in the international exhibitions that followed in the wake of the Great Exhibition of 1851. The show's pavilions became cathedrals of technological progress and arenas for imperial one-upmanship. Telescopes and other instruments played a role in this rhetoric, but it could backfire, as it did at the Great Exhibition. One of the shocks of 1851 was the realization that British optical manufacturers were no longer unchallenged world leaders, with French and German firms showing their evident skills. As Lyon Playfair admitted after the exhibition, British makers were 'justly astonished at seeing most of the foreign countries rapidly approaching and sometimes excelling us in manufactures'.

Large telescopes had a significant role at international exhibitions as technological centrepieces that provided the spectacle visitors came to enjoy. The most dramatic was the largest refracting telescope ever made, the Grande Lunette of the 1900 Exposition Universelle in Paris. As the new century opened, France wished to

40 The tube of the Grande Lunette in the Palais d'Optique of the Paris Exposition Universelle, from *Le Panorama* (Paris, 1900). The journalist Michel Corday described the telescope rather grandly as 'the vertebral column of this vast organism', while admitting that it 'singularly resembles a water conduit'.

show the world that it was flourishing despite recent traumatic decades that included the infamous Dreyfus affair, the Panama scandal and the Franco–Prussian War, 1870–71. As the centerpiece of the Palais d'Optique, the Grande Lunette was to be a structure as impressive as the Eiffel Tower that would push France to the forefront of telescopic astronomy. The 60-m (19 ft 8 in.) long refracting telescope was so large that it could not be pointed to the sky, but lay horizontally with light directed through it from a giant siderostat (a device that compensated for the Earth's rotation to maintain a steady image). The main tube ran along a finely decorated gallery (Fig. 40), with its celestial images projected onto a large screen in, appropriately, the Galileo Room.

Although it was the star of the exhibition, the telescope had many critics, one sourly anticipating beforehand that, 'one will be able to eat dinner in the telescope . . . for it is not so much the love of science as the love of gain that inspires these projects'. It also became a target for jokes and cartoons, the London *Daily Mail* mocking its supposed purpose of revealing the Moon's inhabitants. Away from its detractors, astronomers did use the great telescope to make astronomical drawings and photographs, but it was finally dismantled in 1910. No refracting telescope of its size has been completed since.

The Grande Lunette and, more practically, the Yerkes refractor marked the culmination of a size race in lens-based telescopes, one that saw astronomical telescopes placed in dedicated observatories world-wide. By this time, other telescopic instruments were just as common, in surveying, navigation and other professional activities, as well as in the hands and pockets of all sorts of people. In a century and a half the telescope achieved world-wide domination. It was a dominance that also affected the imagination.

Chapter Seven

THE TELESCOPIC VIEW
1720–1900

"Curiouser and curiouser!" cried Alice (she was so much surprised, that for the moment she quite forgot how to speak good English). "Now I'm opening out like the largest telescope that ever was! Goodbye, feet!"

Lewis Carroll, *Alice's Adventures in Wonderland* (1865)

As telescopes went global, their spread inevitably had an impact on the arts and culture, as a practical tool and as a symbol. But an important question remained. What did one really see through a telescope? In one sense it was already answered. The telescope revealed the truth. It was ideal for astronomical research and as an everyday tool on land and at sea. Yet the possibility that telescopes distorted the truth or distracted the viewer from other more important things continued to be a source of concern. These possibilities shadowed the deployment of telescopic imagery in art and literature, in particular as attitudes towards science waxed and waned. Despite such concerns, the results of telescopic observation were also a source of inspiration in a fertile period of astronomical discovery.

While the telescope was already long established in professions such as seafaring, it was in the eighteenth century that telescopic and other optical aids began to have some impact on artistic practice. By the early years of the nineteenth century, landscape artists had several different instruments to choose from. These included the camera obscura, camera lucida and Cornelius Varley's Graphic Telescope, patented in 1811. As both an artist and optical instrument maker, Varley was just the right person to come up with the idea. His invention was a combination of mirrors and a low-powered telescope that allowed artists to look simultaneously at a view and at the paper on which they were drawing. It had the added advantage of enlarging or reducing the scene to any size. The design came out of Varley's interest in the naturalistic depiction of landscape through the close observation of detail. He and his brother John, also a talented artist as well as a skilled astrologer, used it for portraiture and landscapes (Fig. 41). So did artists such as John Sell Cotman.

A more ambitious use of the Graphic Telescope was by a land surveyor named Thomas Hornor, who, in 1820, was given permission to place a hut onto scaffolding being used for work to the dome of St Paul's Cathedral. From this precarious position, Hornor

41 *Rowing boat at Lowestoft, 1807*, by Cornelius Varley, from *Varley's Book of Boats . . . No. 1*. The book's pictures were drawn with the Graphic Telescope, which Varley believed was 'very valuable for drawing shipping and boats, the various curves of which cannot be known, yet are hereby given quite correct'.

sketched a 360-degree view of London through a Graphic Telescope. Over the months he endured all weathers, only just surviving his 'observatory' being torn from its fastenings in a storm, but still managed to produce 3,800 sheets of drawings. Some years later, they were turned into a painted panorama in the specially built Colosseum in London's Regent's Park in 1829. The panorama was so large that it had to be viewed at several levels, accessed by England's first passenger lift, also designed by Hornor. Popular as it was with the public, the project ruined him financially. Hornor absconded to America later that year.

Optical aids like the Graphic Telescope were not universally popular. Critics argued that they left no room for artistic creativity because they reduced drawing to a mechanical process. Varley responded that, 'instruments not being masters, they cannot make

42 'Nelson at Copenhagen', from *Nelson*, a series of cigarette cards produced for Wills's Cigarettes, 1905. Nelson 'looking' through a telescope with his blind eye was a well established image one hundred years after his death at the Battle of Trafalgar.

artists of those who want the necessary previous knowledge and practice', although his instrument would 'greatly facilitate the progress of an artist'. His thoughts echoed the words of a Japanese painter, Maruyama Ōkyo, who recommended optical aids to improve drawing, suggesting in 1768 that aspiring artists, 'Fix a reflection in a mirror, or else focus on an object with a telescope, and copy what you see down'.

As an artistic subject, telescopes were still common as personal accessories and as symbols of the rank and profession of naval officers in portraits and other images (Fig. 31). Many played it straight, but a more unusual reading came from one of the stories of the British naval hero Horatio Nelson. It started from an incident

43 *A New Leaf for an Old Book of Common Prayer*, by James Sayers, 1807. The Marquess of Buckingham, as the new Guy Fawkes, advances towards the House of Commons, but is revealed in a beam from King George III's spyglass. With the King's perception enhanced to supernatural levels, the image turns the usual short-sighted picture of the monarch on its head (see Fig. 30).

at the Battle of Copenhagen in 1801, when the commander of the British fleet signalled to his ships to discontinue the action. It was later said that Nelson put his telescope to his blind eye (the result of a wound at the Battle of the Nile) and announced, 'I really do not see the signal' (Fig. 42). Ignoring the order, Nelson ensured a decisive British victory by continuing the fighting. Whether or not it happened, this became an iconic tale in the Nelson legend that flourished in the nineteenth century. It also changed over time: Nelson was later said to have remarked, 'I see no ships'. By deliberately ignoring what his telescope revealed, the myth suggested, Nelson showed his strategic vision. It fitted neatly with the idea that Nelson's insubordination was one of his strengths.

Other images explored the idea of enhanced vision even more playfully, perhaps extending the telescope's abilities to supernatural levels (Figs 2, 43). But this was just one side of the coin. Some satirical treatments played on more salacious interpretations. Telescopes became voyeurs' tools, with men in the theatre peeping up a ballerina's costume, or the instrument's obvious phallic likeness was exploited (Fig. 44). Such sexual imagery was not restricted to English satirical prints. A number of Japanese *ukiyo-e* (pictures of the floating world) played on the erotic potential of the instrument's gaze and on its shape, suggestively fingered by a young man or woman. More bawdily, *biidoro*, a Japanese word for an extending telescope was also slang for penis. A short poem from 1753 saucily described,

> The telescope –
> Such over-confidence
> Then the eyes go back to how they were.

There were more sober uses. Telescopes became increasingly common as props indicating scientific interest and learning. During the eighteenth century science became firmly embedded in English and European society as the new theories of Isaac Newton and his followers were popularized in books, journals and magazines and public lectures. But it was not a smooth ride. There were still those who remained suspicious of science and disputed its usefulness. As a prominent scientific instrument, the telescope was bound up in these debates. It could be used as a tool for promoting the new science but also as a symbol of all that might be wrong with it.

44 *Progress of Gallantry or Stolen Kisses Sweetest*, by Thomas Rowlandson, 1814. Rowlandson plays on the voyeurism of telescopic observation and the instrument's phallic potential, heavily emphasized by the placing of the gun barrel.

The move to spread scientific knowledge beyond small groups meeting in institutions like the Royal Society began in the second half of the seventeenth century. With Newton as a national hero by the early 1700s, increasing numbers of publications and lectures began to spread the word. These all sought to present scientific knowledge as useful and appropriate for men and women of fashion. In other words, they sought to distance science from its characterization as an irrelevant activity of interest only to solitary, ill-mannered and bookish fools – the stock figures of seventeenth-century satire. Telescopes and other scientific instruments were now presented as tools of enlightenment helping readers to understand God's universe.

To emphasize the social rather than solitary aspect of the new science, many works took the form of a dialogue or a conversation, following the model of Bernard le Bovier de Fontenelle's influential *Entretiens sur la Pluralité des Mondes* ('Conversations on the Plurality of Worlds') of 1686. Translated into English several times over the next decades, Fontenelle's dialogue had many imitators in the eighteenth century. Like Fontenelle, they promoted the acquisition of 'useful and real Learning' for personal improvement. Often aimed at women, they showed science as a family pursuit that was ideal for teaching morals and manners, aided by the use of fine instruments.

Appropriately, one of the most popular was *The Newtonian System of Philosophy; Adapted to the Capacities of Young Gentlemen and Ladies* by 'Tom Telescope'. Probably written by John Newbery, it was first published in 1761 and reprinted many times into the nineteenth century. Tom Telescope is the narrator of a series of lectures and discussions about subjects ranging from the solar system to the human senses. Some take place in the Marquess of Setstar's observatory, where Tom explains how to use telescopes to explore the universe and explain the latest scientific theories. In doing so, he points out, his students are revealing God's universe. Tom's telescopes are just as much moral and religious instruments as scientific ones.

45 Frontispiece to Benjamin Martin, *The Young Gentleman and Lady's Philosophy* (London, 1759). The wealthy domestic interior has a telescope and celestial globe placed prominently as beautiful possessions and tools of learning.

Other texts followed the same pattern. Telescopes appear prominently in Benjamin Martin's *The Young Gentleman and Lady's Philosophy* (1759), a dialogue between Cleonicus and his sister Euphrosyne (Fig. 45). Exploring the universe with the aid of scientific instruments strengthens religious faith, Martin writes, and can improve a young woman's chances in the marriage market. Euphrosyne initially confesses to being embarrassed that she might appear too masculine by speaking knowledgeably on scientific matters; Cleonicus reassures her that nothing could be further from the truth. A young woman named Euprepia, he tells his sister, has been educated in the sciences and is now 'admired, esteemed and beloved by all Gentlemen of Discernment'. This was a great change from a generation earlier, when James Miller's play *The Humours of*

Oxford (1730) announced that, 'a woman makes as ridiculous a Figure, poring over Globes, or thro' a Telescope, as a Man would with a Pair of Preservers mending Lace.'

Martin's and other works tried to show that learning science was a social activity carried out through conversation, helped by the use of instruments. It was, they suggested, a safe domestic activity for anyone who aspired to be 'polite'. This was a context in which owning telescopes and other instruments was as much about presenting the right image, that of the gentleman or gentlewoman of learning, as gaining knowledge. Owning a fine telescope made you a fine person. It was no coincidence that Martin and other authors sold telescopes and other scientific instruments.

The idea that the telescope revealed moral and spiritual, as well as scientific, truths continued through the nineteenth century. Towards its end, the popular astronomy author Agnes Mary Clerke wrote that observing the heavens leads 'towards a fuller understanding of the manifold works which have in all ages irresistibly spoken to man of the glory of God'. A few years later she even discussed how the spectroscope and camera as additions to telescopes were revealing the complexity of a divinely designed universe. Wherever the telescope was pointed, it revealed the same divine pattern of design as on Earth.

Where there were promoters of science, there were detractors. Often its critics ran with the charges levelled against scientific practitioners in the seventeenth century. The renowned satirist Jonathan Swift attacked science and other modern learning on the grounds that it diverted attention from the more important task of striving to live piously and virtuously. A telescope was an appropriate metaphor in this context. In the third voyage of *Gulliver's Travels* (1726), Swift describes the Grand Academy in the Laputan city of Lagado, an institution full of absent-minded philosophical speculators that lampoons the Royal Society of London. The powerful telescopes with which the Laputans spend 'the greatest Part of their Lives in observing the celestial Bodies' emphasize their detachment from the real world. Elsewhere, Swift uses telescopic imagery more sympathetically, as Gulliver encounters beings who are twelve times smaller (the Lilliputians) and twelve times larger (the Brobdingnagians) than himself. The correlation with the magnifying and diminishing power of Gulliver's 'pocket perspective' reinforces the idea of rescaling. A century and a half later Lewis Carroll used the same imagery as Alice shrinks and grows in *Alice's Adventures in Wonderland* (1865), but with the added similarity to the extension and contraction of a collapsible telescope.

Swift's point that telescopes typified the way in which scientists overlook the essential hailed back to the seventeenth-century image of the virtuoso as an isolated seeker of useless knowledge. It was an image that persisted, even though Martin and others were portraying scientific learning as a social and virtuous activity. Satirical artists like Thomas Rowlandson also replayed the stereotype, with an observer's gaze distracted skywards from more immediate events (Fig. 46), while images of King George III with a spyglass for his poor sight emphasized his ignorance of worldly affairs (see Fig. 30).

Rowlandson's vision of telescopic distraction was lighthearted, but those more completely disenchanted with science viewed the telescope as a sinister device that produced unnatural visions. They believed that scientific discoveries made with telescopes and other instruments were killing off humanity's relationship with God and nature. For Romantic writers, the new science and its mechanistic explanations limited the universe to separate, measurable entities. It constrained humanity, denying individuality and the possibility of understanding the universe through imagination and emotional experience. Some authors sought alternative worldviews that tried to bring science and spirit together, others attacked science itself.

In E.T.A. Hoffmann's macabre tale, *Der Sandmann* (The Sandman, 1816), Nathanael is bewitched by a telescope given him by the evil Dr Coppelius, disguised as a seller of 'glasses' (meaning

46 *Looking at the Comet till You Get a Criek in the Neck*, by Thomas Rowlandson, 1811. Although it draws on the recent appearance of a comet, the satirical message was rather old. Distracted by looking through his telescope, the old man fails to notice the amorous encounter between the younger woman, perhaps his wife or daughter, and a young man.

telescopes and barometers). Confused by the telescope's distorted reality, Nathanael falls in love with an automaton, abandoning the woman he has loved for many years. At the end, he throws himself to his death. In Hoffmann's disenchanted vision, scientific instruments create false images of the world, destroying humanity's emotional and spiritual connection to all around it. It was a view shared by W.B. Yeats, whose 'Song of the Happy Shepherd' warned readers to seek,

> No learning from the starry men,
> Who follow with the optic glass
> The whirling ways of stars that pass . . .

For Yeats, Hoffmann and others the truths that scientists revealed through their instruments had nothing to do with being human.

Others engaged more positively with the new scientific findings. In astronomy, the telescope was revealing the tremendous size of the universe and the extraordinary number of stars it contained. William Herschel estimated that a single field of view of his 'large' 20-ft (6.1 m) telescope contained as many as 50,000 stars and that a larger instrument would reveal still more. Inevitably, this was something that gripped the imagination. Thomas Hardy wrote that he composed *Two on a Tower* (1882), 'to set the emotional history of two infinitesimal lives against the stupendous background of the stellar universe'. Lady Constantine, the story's heroine, falls in love with a young man who is passionate about astronomy, helping him to build an observatory on top of a family monument. The young astronomer uses the new telescope to show her a universe of overwhelming size, with all 'its beauty and its frightfulness', the universe of which Herschel wrote a century earlier. While the central theme is humanity's rescaling in the face of this new universe, there is an elegant reversal of the traditional image of the virtuoso/scientist. In contrast to these foolish old men ignorant of all around them, the youthful astronomer's ability to be distracted from earthly things is part of his allure. As Lady Constantine watches him observing through the telescope, she realizes that, 'those were eyes which habitually gazed, not into the depths of other eyes, but into other worlds'.

Miriam's Schooling (1890), a short story by William Hale White, also draws on the power of telescopic revelations, as Miriam gains a new appreciation for her seemingly modest situation through a deeper understanding of the universe. The turning point occurs as she looks through a telescope while the nature of the heavens is explained:

> The stars had passed like this before her eyes ever since she had been born, but what was so familiar had never been emphasized or put in a frame, and consequently had never produced its due effect.

Hardy and White use the telescope to think about the universe and about this world in particular. It was the same with works of science fiction, a genre that lapped up the new telescopic discoveries. In Robert Braine's wonderfully titled *Messages from Mars, By the Aid of the Telescope Plant* (1892), a man shipwrecked on an island is allowed to use a powerful telescope that the islanders have created from a plant with lens-like properties that grows there:

> I took my seat before the tube, and yet I hesitated to look. There was a mysterious fascination about it; my breath came and went in gasps, my pulse mounted to unknown heights, I was pale and flushed by turns. Then with a sudden quick resolution, I seized the tube and applied my eye to it – there bathed in a rosy light lay the plains, valleys and mountains of Mars before my eyes.

Braine's story, like H.G. Wells's 'The Crystal Egg' (1899), draws a stark contrast between the ideal Martian societies revealed optically and Earth's social problems. Both stories were about looking not away from Earth, but towards it.

These two works were part of an imaginative shift caused by recent astronomical discoveries. Sixty years earlier, it was still possible to imagine life on the Moon. In August 1835, a New York penny paper, *The Sun*, ran a series of articles reporting startling lunar discoveries supported by lengthy, scientific-sounding explanations. These were made, it said, by the eminent English scientist John Herschel with a telescope 'uniting all the meritorious points in the Gregorian and Newtonian instruments, with the highly interesting achromatic discovery of Dolland [*sic*.]'. He had installed it near Cape Town. One thing was true at least: John Herschel was doing astronomy at the Cape at the time (see Fig. 39).

Over the next few days, *The Sun* reported Herschel's amazing discoveries, each more breathtaking than the last. The Moon was alive with plants, birds and curious animals: a spherical amphibian, bluish creatures like unicorns, and beavers just like those on Earth, apart from their lack of tail, 'invariable habit of walking upon only two feet' and ability to make fire. As if these fantastical descriptions weren't enough, readers learned of a race of winged humanoids living in pastoral harmony near beautiful temples of polished sapphire. These, they read, had been scientifically named 'Vespertilio-homo', or man-bat.

As the reports escalated so did the readership, boosting *The Sun*'s circulation above that of other American papers. It was even reported that a clergyman wished to supply bibles to the Moon's inhabitants. But a few days later, Richard Adams Locke, one of the paper's journalists, revealed that the story was a satire on those who believed in life on other planets and who would accept anything that came with a plausible scientific backing. This had not stopped the story spreading to Europe. John Herschel later complained that 'I have been pestered from all quarters with that ridiculous hoax about the Moon – in English French Italian & German!'

Genuine speculation about the possibility of life elsewhere was not affected by these satirical swipes. What did change was the thinking about where that life might be, since astronomers were becoming certain that the Moon had no atmosphere and no water. Theories about extraterrestrial life shifted to Mars, encouraged by

William Herschel's comments that the red planet has 'a considerable but moderate atmosphere, so that its inhabitants probably enjoy a situation in many respects similar to ours'. By the 1870s, as telescopes were being trained on Mars to monitor its close approach to Earth, writers began exploring the possibilities Herschel raised. Some, like Braine and Wells, used it to think about this world; others made Mars the exotic destination for Earth's explorers in the oceans of space.

These writings were given a further boost in 1877 by the announcement of 'canali' (channels) on the surface of Mars. The discovery came from a good source, Giovanni Schiaperelli, Director of the observatory at Brera in Italy. At first, a few people questioned the idea. Nathaniel Green, a trained artist, saw no such thing through his telescope in Madeira and insisted that other astronomers 'have not drawn what they have seen', but 'turned soft and indefinite pieces of shading into clear, sharp lines'. Green's criticisms aside, as Schiaperelli's published drawings became sharper and more regular in appearance, more and more astronomers began to recognize the canali for themselves. Just as Galileo's first results helped observers see the Moon and other heavenly bodies differently, so Schiaperelli's drawings influenced how other astronomers interpreted their observations. They saw what they expected to see, and what they expected to see were the canali. By the 1890s most observers believed that there were regular structures on Mars and the idea took hold that they were not just channels but canals carrying water (Fig. 47). Schiaperelli was hailed as the 'Columbus of Mars'.

An obvious conclusion was that the canals had been created by Martians. The idea's most ardent supporter was Percival Lowell, a wealthy businessman with his own observatory at Flagstaff, Arizona. After first observing Mars in 1894, Lowell published extensively on Mars and its canals, even in the face of growing criticism, as other astronomers began to conclude that the canals did not, after all, exist.

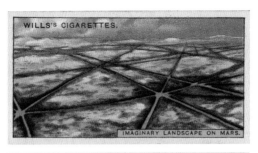

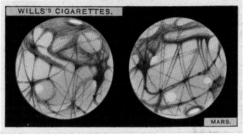

47 'An Imaginary Landscape on Mars' and 'Two Views of Mars', from *Romance of the Heavens*, a series of cigarette cards produced for Wills's Cigarettes, 1928. The 'Imaginary Landscape' depicts the canals as waterways built by Martians. By this time Percival Lowell's theories about intelligent life on Mars were discredited, but the card's commentary still says that 'vegetable life seems probable'.

Serious doubts about the canals' existence came from E. Walter Maunder at the Royal Observatory in Greenwich, who suggested that they might be a creation of the mind. Maunder and Joseph Evans later carried out experiments with boys from the nearby Royal Hospital School to show that people instinctively join up random dots into distinct features. This was proof enough for many, although Lowell dismissed Maunder's 'small boy theory', insisting that only telescopic observation could settle the matter. He even published photographs from his observatory to support his claims. While the scientific community began to dismiss the canals, Lowell remained a firm believer to his dying days and had a huge public following.

While astronomers wrangled over what they saw on Mars, many writers became excited by the possibilities these new observations suggested. The most famous was H.G. Wells in *The War of the Worlds* (1898), which in its opening lines turned the idea of telescopic viewing back onto Earth:

> No one would have believed in the last years of the nineteenth century that this world was being watched keenly and closely by intelligences greater than man's and yet as mortal as his own; that as men busied themselves about their various concerns they were scrutinised and studied, perhaps almost as narrowly as a man with a microscope might scrutinise the transient creatures that swarm and multiply in a drop of water.

Scrutiny, of others and ourselves, is a recurring theme. At the book's end, with the Martian invasion defeated by the very small, Earth's bacteria, the observers have swapped places: Earth's telescopes scan the skies in 'eager scrutiny of the Martian disc', watching out in case of another attack.

By the close of the nineteenth century, science was on a sure and certain footing and the telescope, one of its iconic instruments, was revealing a vast universe and new worlds that captured the public and literary imagination. Debates still raged. Did telescopes deceive or were their revelations real? What did they really show? These questions had shadowed the telescope since its invention and did not seem likely to go away, even as new devices large and small were created in the century to come.

Chapter Eight

TELESCOPES, BINOCULARS AND MODERN LIFE

Sitting around looking out of the window to kill time is one thing, but doing it the way you are with binoculars and wild opinions . . . is diseased!

Rear Window (dir. Alfred Hitchcock, 1954)

While the telescope evolved in fits and starts from the moment of its introduction, it was in the twentieth century that some of its early promise was finally fulfilled. Above all, its development as a hand-held instrument saw binocular devices become the standard for all sorts of uses, overt and covert. At the same time, continuous technical advances have brought telescopes and binoculars to many new types of user.

An instrument for both eyes was the natural development of something that extended the power of seeing. So it is no surprise that one of the first things the Dutch States-General asked Hans Lipperhey to do was to produce a binocular version of the instrument he presented in 1608; within a couple of months he produced one with two of his telescopes mounted side by side. In the centuries that followed, a number of people made similar twin-bodied telescopes, and by the 1820s the German firm of Voigtländer was selling binocular opera glasses in quite large numbers. But the limited field of view and low magnification of these Galilean binoculars limited their use, one British officer dismissing the field glasses issued during the Crimean War (1853–56) as 'useless toys'.

An important breakthrough came in 1854 when a former Italian artillery officer and military surveyor, Ignazio Porro, patented a new image-erecting system. This used prisms that effectively folded up the light path, allowing him to make compact devices with a good field of view but without draw tubes. Porro was an instrument maker in Paris by this time and mainly produced monocular devices, including the 'Longue Vue Cornet' and 'Lunette Napoleon III', given its name after he presented two to the French Emperor in 1855. Porro intended his instruments primarily for military use, since they were quicker to use than telescopes with several draw tubes. As a magazine noted, he knew 'how precious are the moments spent in front of the enemy'.

Porro's ideas did not catch on immediately but were taken up more fully towards the end of the century. Once German firms like

48 *Things as they are, 1917. Things as they were, 1887.* This gently mocking image records some of the changes in the Royal Navy over the turn of the century, including the move from telescopes to binoculars (in the leather case worn by the 1917 officer).

Zeiss overcame the difficulties of producing good prisms relatively cheaply, they began to make binocular devices in large numbers. These new prismatic binoculars became extremely popular and rapidly replaced telescopes for many uses (Fig. 48).

As an example, the British army became painfully aware of the inadequacy of Galilean binoculars compared to the prismatic models its opponents were using during the Boer War (1899–1902), and introduced prismatic binoculars in 1907. By the opening year of the First World War in 1914, these had become indispensable, with over 58,000 prismatic binoculars ordered, compared to just 14,000 telescopes and less than 6,000 Galilean binoculars.

But the telescope did not entirely fall from use. Although binoculars were generally more suitable at sea, many officers continued to use telescopes, which often had greater power. Describing Admiral Jellicoe, Commander-in-Chief of the Home Fleet in 1915, *The Times* noted, 'A telescope under his arm, too, as he received his guests. One liked that. He keeps watch over that Fleet himself when he is on the quarter-deck.'

In the same period, the telescope and its offshoots came to fulfil its early potential in a second way: as the strategic military device of which Lipperhey, Galileo and others had boasted. As the twentieth century opened, the importance of optical munitions was becoming increasingly clear as tensions between Britain and Germany heightened. As an arms race escalated, their optical industries covertly developed new seeing aids for use in the anticipated conflict. The British Royal Navy had first introduced telescopic sights for its heavy guns in 1880, but by the early 1900s larger guns with increased firing ranges needed better gun-laying and sighting devices. They needed more sophisticated distance-measuring devices too, such as the optical rangefinder. The introduction of the submarine also required a new seeing instrument, the periscope.

The new range of telescopic and binocular instruments for the military included stereo-telescopes, field periscopes, trench binoculars and panoramic telescopes, with field glasses always needed in huge numbers. The military importance of these became obvious in both 1914 and 1939, when surges in demand for equipment stretched optical manufacturers to their limits and beyond. During the First World War, the British government was forced to build new factories, expand existing workshops and install modern machinery to increase the production of optical equipment. Meanwhile, advertisements in newspapers and journals urged the public to send their telescopes, binoculars and other optical instruments to the War Office or the Admiralty.

It was the same story two decades later as the outbreak of the Second World War saw a new surge in demand from the armies and

49 Prismatic binoculars by Research Enterprises Ltd, Toronto, Canada, c. 1945. Set up in 1940 to meet Canada's wartime needs for optical instruments, REL began production the following year. The company produced c. 25,000 of these 7 × 50 binoculars.

navies of the warring nations. As H.C. McKay wrote, the glass lens was still 'the giant's eye of modern mechanized warfare, without which armies would be practically blind'. As the world's weapons grew increasingly complex and powerful, much of the fighting was taking place at distances from which combatants could no longer see each other: artillery fire was directed by officers viewing targets through binoculars, telephoning instructions back to their guns miles away. Such powerful weapons needed new high-precision optical instruments, such as dial sight directors, but binoculars and other basic equipment were still needed in huge quantities.

Again, the increased demand put a strain on suppliers. Germany, Italy, Japan, Britain, France and America had optical industries initially able to meet the new need. Other countries, like Australia and Canada, had to create new industries more or less from scratch to make the sighting and other telescopes, gun and rifle sights, periscopes, rangefinders and binoculars they needed (Fig. 49). Even countries with established industries had to take further steps, Britain forming Umbroc, a secret company producing optical-quality glass; its name was a combination of 'umbra' (shadow) and 'optic'.

The volume of demand receded after the Second World War, but optical munitions are still a key military technology. They now include high-powered telescopic sights for hand-held weapons, night vision optics and image enhancing devices, but binoculars are just as essential as ever. Other professional activities have also seen an increasing sophistication in the optical technologies used, including surveying, where the move to satellite mapping has led to the development of theodolites with digital components and integrated GPS (Global Positioning System) receivers.

A close relation of the military use of the telescope and its spin-offs is the development of surveillance equipment, both in war and for 'national security'. Optical technologies for surveillance have advanced dramatically in the last hundred years largely as a result of developments in lens systems. Magnifying lenses, the forerunners of telephoto lenses, were devised in 1834 by the English engineer Peter Barlow for use with telescopes. They were applied to cameras in the 1890s and used for aerial reconnaissance in the First and Second World Wars. Since then the ability to capture high-quality pictures from aircraft and satellites has improved dramatically, with publicly accessible media such as Google Earth now hinting at the capabilities of more sophisticated classified systems.

Modern society's increasing obsession with surveillance is also becoming more and more evident in the growing numbers of cameras with telescopic and zoom lenses on our streets. In Britain, this began with a limited system set-up for Queen Elizabeth II's coronation in 1953, with permanent systems installed in London by the 1960s. It was estimated recently that there are over four million surveillance cameras in this country, about one for every fourteen people.

The downside is the suspicion that surveillance has got out of hand. Ridley Scott's thriller *Enemy of the State* (1998) plays on these concerns about the information that government agencies and others can now collect; the title sequence presenting a montage of

50 *As Seen Through a Telescope* (1900). Inter-cut sequences show a man lewdly observing a young couple and his view of them. At the end, the young man punches the voyeur.

images from street cameras and satellites, which lie at the heart of the thriller's plot. As an indication of how times have changed, the film reintroduces Edward 'Brill' Lyle (played by Gene Hackman), originally the central character in Francis Ford Coppola's *The Conversation* (1974). While the earlier film concentrates on audio surveillance, the best technologies then available, Scott's later film draws Brill into a world where optical surveillance has become highly sophisticated and open to abuse.

Yet telescopic surveillance need not be restricted to the security services or to 'official' purposes. Almost since they began, cinema and other popular media picked up on people's constant delight in watching what others are up to. In *As Seen Through a Telescope* (dir. George Albert Smith, 1900), one of a number of early films that explored the visual potential of optical aids, a man observes a young couple across a street. The audience shares his view: the young man's hands caressing the woman's foot and ankle, shown within a circular mask to mimic the telescopic view (Fig. 50).

This obvious link between telescopic devices and voyeurism was most famously explored in Alfred Hitchcock's *Rear Window* (1954), in which James Stewart plays Jeff Jefferies, a photographer who is housebound after breaking his leg. As a result, he starts watching and speculating about the lives of his neighbours, aided by binoculars and a telephoto lens (his 'portable keyhole'). Openly exploring what Jeff's girlfriend calls 'rear window ethics', the plot draws him into a murder-mystery, with Jeff pausing at one point to wonder 'if it's ethical to watch a man with binoculars and a long-focus lens . . . even if you prove that he didn't commit a crime?'

Rear Window plays fairly gently with people's voyeuristic tendencies, but clearly reflects a common fascination with prying into the lives of others, something that urban life encourages. As the *New York Times* reported in 1990,

> everybody is always looking at everybody else, but nobody wants to be caught in the act. This is why, from bay windows in brownstones to penthouses so high that helicopters flutter beneath them, there are viewfinders being fiddled with, focusing knobs being twirled and powerful lenses peering deeply into unsuspecting people's eyes.

But optically enhanced people-watching can have a more unsettling side. The telescope, some authors and film-makers suggest, may allow one to look at others, but not to connect with them. In 'Wakefield, 7E', one of the linked short stories in Gabriel Brownstein's *The Curious Case of Benjamin Button, Apt. 3W* (2002), a former lawyer who walked out of his previous life now lives secretively in Benjamin's block. From his new apartment he watches his unsuspecting wife and daughter still living across the street:

> He stared through a telescope while they cried, skipped meals lost weight and friends . . . While Ada sliced chicken, Wakefield focused binoculars on her knuckly hands. When she took lovers to bed, he gazed at the shade to the window to what had been his bedroom.

Wakefield's optical aids simultaneously offer proximity and isolation from those he has loved, and perhaps still loves.

Whether or not one views such pastimes as sinister, there are other forms of recreational viewing that are harmless and enjoyable,

51 Hawk watching at the Illinois Beach State Park, Zion, Illinois, USA, November 2008.

whether they involve looking at this world or at others. The development of seeing aids has been central to almost all of them. A good example is bird-watching, which grew in popularity in the twentieth century. As early as 1889, Florence Merriam's *Birds through an Opera Glass* highlighted the benefits of optical aids, but until the advent of prismatic binoculars, the equipment available was largely unsatisfactory. As Max Nicholson, an early champion, remarked,

> It is evident to anyone who glances at a party of ornithologists trying to use telescopes . . . that sooner or later either the telescope or the bird-watcher will have to be entirely redesigned. The former would be more convenient.

Nicholson and others naturally warmed to the newly introduced prismatic binoculars, which were lighter and had a good field of vision. In doing so, they changed the nature and role of bird-watching. The use of improved optical aids and recognition guides gradually legitimized sight records as evidence of the arrival of rare birds; before then the birds had to be physically produced (dead) for a sighting to be credited. Amateur bird-watchers also began to take part in surveys and migration studies, for which good instruments were essential. This further encouraged the improvement of specialized equipment, including the telescope and tripod, of which lighter models to suit bird-watchers' specific needs began to appear in the 1970s (Fig. 51). Still, it is binoculars that have become the trademark bird-watching accessory, whether as a badge of office for those engaged in the pastime or in humorous images, including the rather less upstanding 'bird-watchers' of one *Carry On* film.

Bird-watching is just one of many pursuits that commonly use binoculars. Small, powerful and easy to handle, they are ideal for any activity that requires observation from a distance and have long been a regular accessory among horse-racing crowds and at other sporting events (Fig. 52). One historian has even recounted the sad tale of a Sussex vicar forced to give up bird-watching for botany since whenever his parishioners saw him with binoculars they thought he was off to the horse races.

Of the many hobbies with which people now occupy their spare time, amateur astronomy is the one most obviously associated with the telescope. Although its history is considerably older, it was at the end of the nineteenth century that some meaningful differentiation between 'amateur' and 'professional' astronomers began to emerge. As amateur astronomy grew in popularity in the twentieth century, it spawned new publications including *Sky & Telescope* (begun in 1941) and television programmes such as *The Sky at Night*, broadcast in Britain since 1957. It also received a boon from the availability of cheap surplus equipment after the Second World War, while the decades after the first Moon landing in 1969 saw an explosion in the number of amateur observers and the range of products and publications from which they could choose.

52 A couple at the horse races on a postcard of c. 1899. Even before the widespread introduction of prismatic versions, binoculars were commonplace at sporting events.

A significant telescope-making industry has evolved to serve this new market, with firms like the Meade Instruments Corporation (founded in 1972) producing medium-sized telescopes for the non-professional observer. The combination of a growing market and competition between these new firms has in turn led to a reduction in prices and great improvements in the optical technology available, including the introduction of 'Go To' telescopes, which incorporate computer controls able to locate specific celestial objects. First sold in the late 1980s, they made a significant impact through models like the Meade LX200 series, introduced in 1992. As observers testify, these telescopes have revolutionized amateur observing (Fig. 53). The availability of portable, high-quality telescopes at affordable prices has also fed the growth of astronomical tourism,

53 Observers from the Flamsteed Astronomy Society on Blackheath, London, January 2008.

as those eager for new astronomical experiences travel with their equipment to the best viewing spots, perhaps chasing down eclipses as they become visible from different countries.

As well as growing in numbers, amateur astronomers have managed to adopt a range of roles that complement the work of their professional counterparts, even finding areas in which they can make genuine scientific contributions. This might be the collection of data that professional astronomers would not otherwise have the time and resources to gather, which is one important role of the American Association of Variable Star Observers. Alternatively, individual observers might discover new astronomical objects, such as comets, novae and supernovae. An Australian minister, Robert Evans, currently holds the record for the discovery by visual observation of supernovae (forty-two so far), mostly made with a reflecting telescope. This is a skill that requires patience, the right instruments and an excellent memory of the patterns of hundreds of stars in order to pick out new phenomena.

Although telescopes are the iconic astronomical instrument, binoculars have also had a huge impact. Many astronomical writers confirm that binoculars are the most versatile, essential equipment for the amateur observer, with a decent pair recommended in preference to a telescope for those starting out. For a number of experienced astronomers, it was viewing the sky with binoculars that first aroused their interest, even if they have gone on to use larger telescopes. The wide field of view also makes binoculars ideal for scanning large areas of sky when hunting for new celestial objects. So although Robert Evans prefers a telescope for spotting supernovae, the late George Alcock, one of Britain's most famous amateurs, used binoculars for hunting comets and novae. He discovered five of each, including a comet spotted with hand-held binoculars through a double-glazed window at the top of his stairs.

In the end, one should not forget that most amateur astronomers are not really on the lookout for new discoveries, but observe simply for the pleasure it offers. 'By looking through your telescope', David Levy writes, 'you can travel through space as far and as fast as you like.'

Modern society has found many uses for the telescope and its offspring. As tools for surveillance they have been used for pleasure and profit, for open war and covert spying. To many owners, they are also fine objects in their own right: 'It's like having books you never read', a Manhattan telescope seller once observed, 'If you're smart enough to own a globe, a telescope and a set of Shakespeare, you have good taste.'

Above all, it is the sheer diversity of types and uses that have characterized the telescope's last century, something that will no doubt be true as hand-held and mid-sized telescopic equipment is developed in the decades to come. It is a tale of continuous evolution that can also be seen in the development of the telescope's bigger cousins in professional astronomy.

Chapter Nine

MODERN ASTRONOMICAL TELESCOPES

> Watch the skies. Everywhere. Keep looking. Keep watching the skies!

The Thing from Another World (dir. Christian Nyby, 1951)

Although it seemed to reach its natural limits in the size race of the nineteenth century, the astronomical telescope took off in new directions in the twentieth, resulting in instruments of unprecedented size and capability. Sparked by the visionary zeal of a few dedicated individuals and brought to fruition through the hard work of countless others, their power and precision have changed once more the ways in which we view our solar system and the wider universe.

The late nineteenth century saw the refracting telescope developed to its practical culmination in the 40-in. (1.02-m) telescope at Yerkes Observatory. At that time, astronomers in state and university-funded observatories still favoured refractors, which tended to be sturdier and easier to manoeuvre. They also had optical systems that required little maintenance in comparison to the mirrors of reflecting telescopes, which suffered image distortion and the damaging effects of temperature fluctuations. But in the century that followed, new breeds of large reflecting telescope emerged, so much so that for today's astronomers the refractor is obsolete for all but a few specialized uses.

Two developments led to this reversal of fortune. The first stemmed from Léon Foucault's discovery in 1857 of a new technique for silvering glass. This allowed the production of telescope mirrors that were twice as reflective as those made entirely of metal, and which were easy to re-silver if they tarnished. Since one could support the mirrors from behind (which was impossible with a lens), it was now fairly easy to produce very large mirrors capable of gathering much more light than any lens.

The second factor was the growth of astrophysics, based on the photographic and spectroscopic methods recently developed by European and American astronomers. Towards the end of the nineteenth century, a key figure in expanding and promoting this new science was George Ellery Hale, who was to have a pivotal role in the development of telescopic astronomy.

Said to be a charming man with a boyish energy, Hale studied and practised astronomy for a number of years before helping to found the University of Chicago's Yerkes Observatory, of which he became first Director. Although Yerkes was home to the impressive 40-in. refracting telescope, Hale's promotion of astrophysics as a core programme for the Observatory led him to favour reflecting telescopes. It was mirrors, he found, that could gather the large quantities of light needed for the photographic and spectroscopic investigation of very dim celestial objects.

Hale wanted big reflecting telescopes for this work. Luckily, his ability to raise funds from wealthy American individuals and institutions matched his ambitions. This became obvious as early as 1892, when he persuaded the streetcar magnate Charles Yerkes to pay for what became the Yerkes refracting telescope, and again in 1896 when Hale persuaded his own father to buy a 60-in. (1.5-m) diameter mirror from France. The mirror was intended for the Yerkes Observatory as well, but by 1904 Hale had successfully helped to found Mount Wilson Solar Observatory (renamed Mount Wilson Observatory in 1919) with funding from the Carnegie Institution. This new observatory near Pasadena in southern California focused on astrophysics, and was to become the model for astrophysical observatories world-wide. On a good day, one might encounter its Director jogging up the mountain reciting Italian poetry.

Working alongside Hale was George Ritchey, head of instrument construction at Yerkes. Together with Hale, Ritchey designed a telescope for the 60-in. mirror. Unveiled in December 1907, it was the most sophisticated telescope of its day. It also impressed observers. As one later wrote, Saturn and Mars appeared 'as if cut out of paper and pasted onto the background sky', a distinct improvement on the 'muddy or dirty look' they had through a refracting telescope.

Yet even before the 60-in. was completed, Hale had plans for a larger instrument, this time with a 100-in. (2.5-m) mirror. Again, the plans were expensive, but his conviction and enthusiasm paid off as he found scientific supporters and rich benefactors. The first was

John D. Hooker, a wealthy Los Angeles businessman, after whom the telescope was to be named. When Hooker's money ran out, Hale successfully courted the industrialist Andrew Carnegie, whose $10 million donation was to 'repay to the old land some part of the debt we owe them by revealing more clearly than ever to them the new heavens'.

The Hooker telescope was ready in 1917 and remained the world's largest until after the Second World War. Its size and quality soon proved its value and within a few years Edwin Hubble and others used the telescope to redraw the picture of the universe. In October 1923, having photographed the Andromeda nebula with the 100-in., Hubble calculated that it was located almost a million light years beyond our galaxy. The universe, he concluded, was not just a single galaxy, but many galaxies spread across enormous distances. Hubble thus ended the debate over whether the Milky Way marked the full extent of the universe. It did not. More surprisingly, he and his colleagues then went on to show that the universe was not only much bigger than previously thought, but that it was still expanding.

These astonishing revelations were quickly reported in the world's press, accompanied by impressive photographs from Mount Wilson's telescopes. Writers like James Jeans were quick to highlight the philosophical implications. In *The Mysterious Universe* (1930), Jeans explored how Hubble's enlarged universe forced one to reassess humanity's place, explaining that,

> We find the universe terrifying because of its vast meaningless distances, terrifying because of its inconceivably long vistas of time which dwarf human history to the twinkling of an eye, terrifying because of our extreme loneliness.

These ideas soon found their way into contemporary writing. In Olaf Stapledon's *Star Maker* (1937), the narrator undertakes a voyage through a gargantuan universe by projecting his consciousness from a hillside on Earth into the cosmos. There he seeks out and combines with countless alien life forms. The narrative is vast in its conception of time and space as it leads to its culmination, the revelation that the Star Maker has made and discarded many universes and is about to discard this one. For Stapledon, humanity is an infinitesimal part of the cosmos, as Hubble revealed at Mount Wilson and Jeans explained. Yet in the face of this, his novels seek a future for Earth's inhabitants that might avoid the aggression and annihilation that seemed inevitable in the light of reports of the Spanish Civil War. Stapledon's imaginary journey through a rich and complex cosmos is an exploration of other possibilities for humanity.

It was not just space narratives that capitalized on the astronomical revelations from Mount Wilson. Like Stapledon, Virginia Woolf was deeply affected by the suggested scale of the new universe. In *The Waves* (1931), she writes of the Earth as 'a pebble flicked off accidentally from the face of the sun', a mere nothing in 'the whirling abysses of infinite space'. As well as questioning humanity's place in a re-scaled cosmos, Woolf's writings deploy telescopic imagery in more immediate ways. Her short story 'The Searchlight' presents a narrative told at a dinner in a London club. Seeing the searchlight of the title, Mrs Ivimey recounts how her great-grandfather once watched a young couple kiss through his telescope. The tale deliberately deploys the notion of telescopic viewing in a number of ways. The writing mimics the experience of the instrument's restricted field of view – revealing in succession one tree at a time, then birds, then the farmhouse where the couple are seen. It also exploits the idea of telescopes as time machines, since astronomers explained that looking into space was the same as looking back in time. Just as astronomers could look back towards the birth of the universe, so Mrs Ivimey looks back to her own origin: her great-grandfather married the girl he saw in the telescope. But the tale is also about endings and draws on humanity's ephemeral nature that astronomical telescopes revealed, evoked by the threat of impending war and the telescope's military history. Like

54 Brochure for the Mount Wilson Hotel. By the 1930s, Mount Wilson's telescopes were a popular tourist attraction. The location allowed tourists to take advantage of the mountain's walking and skiing opportunities.

Stapledon, Woolf seeks an alternative to the grim possibilities of world conflict.

Within a decade of Hubble's momentous discoveries, Mount Wilson was unchallenged as the world's leading observatory. As the source of startling and thought-provoking revelations about the universe, it drew thousands of visitors, becoming a popular tourist attraction (Fig. 54). One visitor, the screen and stage star Helen Hayes, later recalled her surprise that, 'a tiny bit of glass, which could fit into the socket of a man's eye, could swell out to embrace the whole universe. It seemed to put us in hailing distance of all eternity.'

Sadly, Hale was not a well man by the time Mount Wilson was claiming its place on the world stage. For years he suffered from symptoms including severe headaches, insomnia and the feeling that his mind was whirling out of control, later naming them 'the whirligigus' or 'Americanitis', after Americans' tendency to let ambition drive them to madness. As the symptoms worsened, a ringing in his ears would be followed by the appearance of an elf, who began to give him advice. Despite regular stays at a sanatorium, Hale was forced to step down as Director of Mount Wilson by 1922.

In spite of his illness, Hale kept the title of Director Emeritus and continued to be an influential figure. And even as the Hooker telescope was being completed, he was drawing up plans for a reflector with a 200-in. (5.1-m) mirror. It was a mammoth proposition, but Hale was convinced that the millions of dollars would be well spent, confidently writing in *Harper's Monthly Magazine* that,

> Like buried treasures, the outposts of the universe have beckoned to the adventurous from immemorial times... If the cost of gathering celestial treasure exceeds that of searching for the buried chests of a Morgan or a Flint, the expectation of rich return is surely greater and the route not less attractive.

This time he persuaded the International Education Board (which was endowed by the Rockefeller Foundation) to give $6 million towards the cost of a new telescope and observatory. He also helped to choose the site for the new observatory at Mount Palomar in southern California. He and John Anderson, who took charge of the telescope-building project, toured the mountain in 1934 for a location with dark skies and good seeing. The fern meadow they chose was called *Poharup* (Noise of Falling Water) by the San Luiseño Indians, from the nearby ravine that provided its water source. It sat at an altitude of 5,600 ft (1,700 m).

Although funding was secured in 1928, the project took another twenty years to complete, involving not just building an observatory and telescope, but remodelling Mount Palomar itself. When finished, a 'Highway to the Stars' led up the mountain to its summit, which had been flattened to make way for the 1,000-ton telescope dome. The telescope's design allowed the tube to swing between supports like a large tuning fork, carried on a giant horseshoe

55 Advertisement for Socony-Vacuum, who supplied oil for the Hale telescope, 1948. The telescope's Pyrex mirror is coated with a thin aluminium layer and floats on a bed of compressed oil. According to one astronomer, it was like 'something that had been produced by wizards and elves and set down on earth'.

bearing that floated on a bed of Flying Horse telescope oil (Fig. 55). The sheer size of the instrument, affectionately known as 'Big Eye', meant that astronomers could have the extraordinary experience of sitting in an observing cage at the focus of the 200-in. mirror, riding the telescope through the night while collecting data. Even in this massive project, the romantic image of the lone astronomer exploring the universe with his instrument persisted.

At the telescope's dedication on 3 June 1948, the chairman of the California Institute of Technology's trustees spoke of it as 'the lengthened shadow of man at his best', while the president of the Rockefeller Foundation considered it an instrument to heal an ailing world. The huge telescope, appropriately named after Hale, was then swung north and south, east and west, 'as smoothly as if it were a veteran of the skies'. With observations beginning in earnest the following year, the Hale telescope was soon helping to maintain the USA's preeminence in telescopic astronomy. Mount Palomar stood as a potent symbol of the ambitions and achievements of modern science. Like the Hooker before it, the Hale telescope also helped to resize the universe once more, when Walter Baade used it to show that stars in the Andromeda galaxy were even further away than Hubble calculated.

The Hale telescope was also of lasting significance in that its revolutionary design formed the template for large astronomical instruments world-wide. Even in the 1970s and 1980s, a period in which several telescopes with apertures in the 4-m (157-in.) range were built around the world, most were based on the design at Mount Palomar, including the 4-m Mayall Telescope at Kitt Peak Observatory, Arizona and the 3.6-m (142-in.) European Southern Observatory telescope at La Silla, Chile.

In the last two decades of the century, a new generation of reflecting telescopes has moved away from the Hale model of a single giant mirror on an equatorial mounting. The altazimuth mounting (allowing up-and-down and side-to-side movements) came back into fashion with the building of the 6.05-m (238-in.) Bolshoi Teleskop Azimutal'ny, completed in 1976 on Mount Pastukhov in the north Caucasus in Russia. It was the weight of its large single mirror that favoured an altazimuth mounting, since this was simpler and cheaper to build, although it required better control systems. Luckily, the computer revolution has made this a straightforward proposition. Although the Russian telescope has not proved very successful due to the atmospheric disturbance, it set the model for the mountings of future telescopes.

As well as a move back to altazimuth mountings, the quest for ever bigger light-gathering areas finally reached a point where producing single mirrors presented such overwhelming practical problems that they were becoming prohibitively expensive, even

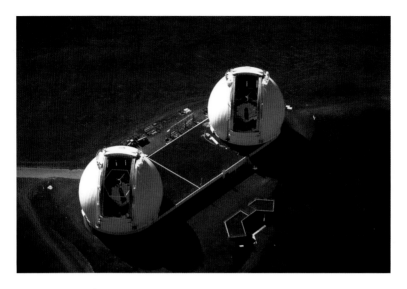

56 The twin telescopes of the Keck Observatory on Mauna Kea, Hawaii. Each of the telescopes, which began observations in 1993 and 1996 respectively, has mirrors made of 36 hexagonal segments.

with the size of budgets now in play. The solution has been to build mirrors in several sections, as was done for the twin Keck telescopes on Mauna Kea, Hawaii. These have 10-m (394-in.) mirrors made of interlocking hexagonal segments, with sophisticated controls to keep the sections aligned (Fig. 56).

Taking a step back from the impressive grandeur of these giants, it is important to remember that the development of the astronomical telescope in the last century has not just been in terms of physical size. A range of smaller-scale technical developments has also helped to reshape the ways in which optical telescopes have been deployed and operated.

In astrophotography, the most significant of these was a new form of reflector, which was devised by Bernhard Schmidt and is still widely used today. Born on an island off the coast of Estonia, Schmidt was a distinctive, solitary figure, having lost his right arm after an accident with a pipe bomb he made as a child. Warned off such explosive hobbies, he began grinding lenses, and after moving to Germany at the turn of the century he set up a small business from a disused bowling alley, where he made telescopes for amateur astronomers. By all accounts he was largely fuelled by brandy, cigars, coffee and cake. After the First World War, when he was imprisoned because he was believed to be using his telescopes to send secret signals to the Russians, Schmidt found himself under the wing of Richard Schorr, Director of Hamburg Observatory. Schorr set him up as mirror-maker at the observatory's branch at Bergedorf. From then on Schmidt worked in secret, remarking that if he were to reveal his methods, 'it would so shock the astronomers and the opticians that I'd probably never get another order'.

It was here that he came up with a design for a telescope with a wide field that did not suffer from spherical aberration, astigmatism or coma, an optical defect that produces stellar images with tails like comets. With the support of the astronomer Walter Baade, who worked with Schmidt in Germany before taking a post at Mount Wilson Observatory in 1931, the new design spread rapidly in the USA in the 1930s. Indeed, the very first telescope to operate at Mount Palomar was a small Schmidt reflector, completed in 1936, while a much larger one (the Oschin Schmidt) was built to operate alongside the Hale telescope, picking out promising targets for Big Eye to observe in detail. It was in wide-angle imaging and photography that Schmidt telescopes showed their quality.

Astronomical telescopes have changed in other ways as a result of modern advances in technology. In the first half of the twentieth century, the photographic plate largely replaced the observer's eye as the primary observing tool, revealing star systems that could never be visible to the eye alone. The development of the photoelectric cell further helped to reduce the need for direct observation, while the advent of silicon charge-coupled devices (CCDs) has marked a huge leap in the ability to gather light by artificial means: while photographic plates capture just one per cent of the available light, CCDs can collect eighty per cent or more. Unsurprisingly, they are now commonly used on spy satellites, including the American KH-11 (or Keyhole-11) reconnaissance satellites, said to be able to resolve objects 2–4 in. (5–10 cm) across at a distance of over 100 miles (161 km).

These developments have increased almost unimaginably the amount of data modern telescopic instruments can gather, while the advent of computers has made processing such volumes of information possible. The overall effect has been to change the way professional astronomers use their telescopes. Instead of the astronomer working intimately with the telescope each night, they are now more likely to sit remotely, perhaps in another country, and rely on an array of intermediate devices and technical staff to collect their results. This has been a disconcerting shift for many of those who have lived through this change: as early as 1971 one astronomer was only half joking when he worried that the telescope was becoming a 'big computer with a large optical analog-input at its periphery'.

In tandem with the computer revolution, other important advances have come from the development of military observing technologies later released for civilian uses. This has led to spectacular advances in imaging, although always a step behind the military. As an astronomer remarked in 1977, 'one of the most infuriating things is to know that a detector far better . . . is orbiting the Earth, attached to a large optical telescope, looking downwards.'

The most impressive of these developments was the introduction of adaptive optics, which successfully eliminates the effects of atmospheric disturbance. This is a problem for all ground-based telescopes, since perturbations in the atmosphere alter the light received (hence the twinkling of stars) and blur the telescope's image. Using elaborate sensors, fast computers and 'rubber mirrors' that are reshaped hundreds of times a second by piezoelectric crystals, adaptive optics produce images that are almost as sharp as if there were no atmosphere. The idea was first proposed in the 1950s

57 The Hubble Space Telescope in orbit above the Earth. Hubble is simply a tube with a mirror at each end: a Cassegrainian reflecting telescope with a 2.4-m (94-in.) primary mirror.

and successfully developed in the 1980s for military satellite imaging. It is only since the end of the Cold War that adaptive optics has become a feature of large astronomical telescopes.

Another key shift has been in the way in which the large telescopes of the last few decades have been funded and developed. Whereas Hale was able to approach philanthropic individuals and funding institutions knowing that they could wholly or substantially fund his proposals, the costs of more recent proposals have been so high that they require national or international funding, and can rarely be the responsibility of single bodies. This change also reflects the way in which astronomy generally has become an internationally collaborative rather than nationally focused endeavour.

The ways in which these new funding models pan out can be seen in the long history of the Hubble Space Telescope (Fig. 57).

Launched into orbit in 1990, Hubble took over forty years to get off the ground (literally). The story began in 1946, when Lyman Spitzer Jr, who had worked on the development of sonar during the Second World War, was put on a US military think-tank on the future of research. His report on the 'Astronomical Advantages of an Extra-Terrestrial Observatory' proposed an observatory in space. This was a bold suggestion, since not even a rocket capsule had been placed in orbit and almost all astronomy was still done from the ground.

Appointed as chairman of Princeton's astrophysical sciences department in 1947, Spitzer spent the rest of his career at the university, all the while working to make his vision a reality by persuading the scientific community and Congress of the project's value. With his characteristic discipline, diligence and politeness, he eventually won through.

The reasons for putting a telescope into space were obvious: avoiding the problems of atmospheric disturbance and light pollution suffered by ground-based telescopes (even those on mountain tops); and allowing astronomers to observe radiation such as X-rays and ultraviolet that does not reach Earth's surface. But as Spitzer pointed out the real contribution would be,

> not to supplement our present ideas of the universe we live in, but rather to uncover new phenomena not yet imagined, and perhaps modify profoundly our basic concepts of space and time.

As with all major projects, the true scientific value, like Hale's 'celestial treasures', would only be revealed once the telescope was up and running. Funding such a project was an act of faith, even if that faith was not entirely blind. Yet although it was proposed at a time when large-scale federal funding of American science made such ambitious projects conceivable, Spitzer cannot have appreciated the final cost of his vision, with current estimates in the region of $5–6 billion.

The big breakthrough came in 1975 when NASA and the European Space Agency began developing the space telescope in earnest, with the US Congress approving funding two years later. Like other major telescope-building projects, the logistics were enormously complex and drew together institutions, groups and individuals from around the world: astronomers, systems engineers, managers, mirror-makers, electrical engineers, computer programmers and countless others. As a project, it went way beyond the scope of Hale's great telescopes, which were still primarily under the control of astronomers. They were now just one of the groups involved and feeding into the project, but were not controlling it. Reflecting this same trend, the astronomers who now use the telescope's data are physically and managerially separated from its running.

The space telescope was also constantly modified in response to shifting and often competing political, economic and scientific agendas. Cuts in NASA's budget meant redefining the scope of the project while attempting to preserve the telescope's scientific credibility to retain the support of the astronomical community, many of whom remained sceptical that such huge sums were being spent appropriately. Between 1973 and 1977 the 'Large Space Telescope' with a 3-m (118-in.) mirror became simply the 'Space Telescope' with a 2.4-m (94-in.) mirror and fewer instruments. Even so, the project remained under constant scrutiny from stern critics: in 1984, when NASA's James Biggs emphasized that 'The Space Telescope will be the eighth wonder of the world', the American politician Edward P. Boland sourly responded that, 'It ought to be at that price.'

In this political landscape and in view of its cost, the telescope's early optical failures were a political disaster and a public embarrassment, with one American senator christening Hubble a 'techno-turkey'. The optical defect, caused by the primary mirror being the wrong shape, was put right by installing a correcting mirror during a 1993 repair mission, and since then Hubble has come to be seen as a success, with widely published images helping to seize the public imagination. Hubble has confirmed the existence of black

58 The Hubble Ultra Deep Field. This photograph was taken by pointing Hubble at a patch of sky that looked empty, in order to see as far as possible. It reveals galaxies as they were over 13 billion years ago when the universe was only 800 million years old.

holes, revealed new galaxies of all shapes and sizes (Fig. 58), and shown us some of the universe's most dramatic moments: a cometary collision with Jupiter and the catastrophic explosions of dying stars. Without doubt, it is now the world's most popular scientific instrument. But the Hubble Space Telescope is not alone up there. It has since been joined by other telescopes, each taking advantage of the observational opportunities away from the Earth's atmosphere, such as the Chandra X-ray Observatory, launched in 1999, and the Spitzer Space Telescope, launched in 2003 to study the universe at infrared wavelengths.

Technologically, Hubble is a straightforward optical telescope. But over the decades in which it came to fruition, new forms of telescope were being developed that broadened the concept of what a telescope is and extended the idea of 'far-seeing' to include wavelengths that the eye could never detect.

This other story began in the 1930s, when Karl Jansky from Bell Telephone Laboratories in New Jersey, USA, was trying to explain a hissing noise suffered by the transatlantic radio telephone service. To help his investigation, Jansky mounted an array of aerials onto the wheels of a Model T Ford. Rotating his new contraption and listening carefully he found that the hissing was the result of interference coming not from Earth but from the stars.

The surprising discovery of an extraterrestrial radio source made the front page of *The New York Times*, but Jansky was moved on to other projects by his employers and only one other person showed much real interest, a radio engineer named Grote Reber. Picking up Jansky's work, Reber built himself a parabolic antenna in his backyard in the Chicago suburb of Wheaton, Illinois, and over the next six years produced the fist radio map of the Milky Way.

Reber was unusual. Other interest was minimal until the development of radar during the Second World War provided a new reason to study extraterrestrial radio sources, since emissions from the Sun were being mistaken for enemy signals. Some work went into the analysis of these sources. This was continued more systematically after the war using surplus radio equipment as the basis of a new discipline, radio astronomy. Over the next decade, ex-military workers recruited into universities in Britain, Australia, the Netherlands and America created a new research community and began to identify extraterrestrial radio sources that were invisible to optical telescopes.

Manchester and Cambridge Universities emerged as British centres for this 'new astronomy', with each developing different forms of instrument. The key figure in Manchester proved to be Bernard Lovell, one of the new breed of radio astronomers fresh from wartime work on radar. Back in Manchester, Lovell began studying cosmic rays with ex-military equipment including a gun-laying radar, moving it to university land at Jodrell Bank south of

Manchester to escape interference from the city's trams. After promising work with modest radio telescopes, he began promoting the idea of a large steerable radio dish, for which he had substantial plans by 1951.

Like other projects of the post-war period, this was a financially ambitious task for which Lovell had to court funding bodies and academic peers. The UK astronomy community wished to re-establish Britain as a world leader in an area where America had taken a lead in the twentieth century, in particular through Hale's great telescopes. The new field of radio astronomy offered a hope of redressing the balance. A 1957 press release placed Jodrell Bank as the next step in a historical sequence of national achievement that included William Herschel and the Earl of Rosse, but which was 'doomed to capitulation to those living in a more favourable climate'. Now, thankfully,

> the devices of war have been transformed into a revolutionary method for the exploration of space, independent of cloud or fog... Britain can once more compete without handicap.

Patrick Blackett, Director of the Manchester University Physical Laboratory, even wryly observed that Jodrell Bank, in one of England's wettest regions, was particularly appropriate 'with its weather, for the study of radio or "blind" astronomy!'

For the British government, whose Department of Scientific and Industrial Research was heavily funding the project, the Jodrell Bank radio telescope was a vehicle of national pride for a country still recovering from the war. Aiming to help national and economic recovery, the Foreign Office funded a film, *The Inquisitive Giant* (1957), which used the telescope to promote British technical expertise overseas.

The radio telescope's military origins in radar also suited it to the rhetoric of the Cold War, with the 1957 launch of the Russian satellite *Sputnik* raising the stakes and allowing Lovell to lobby the Prime Minister, Harold Macmillan. The telescope, Lovell

"Jolly demoralizing tracking other people's satellites AND on something that isn't yet paid for."

59 *Punch*, 14 January 1959. The Jodrell Bank radio telescope was promoted as a symbol of national pride. Its financial problems and the knowledge that other countries had overtaken Britain technologically fuelled more cynical responses.

pointed out, 'would be the most powerful radar equipment in the Western world and would be particularly suited to the tracking of satellites'. These manoeuvres were successful in securing further funding from British military sources, but escalating costs left the project deep in debt, something that became public as it started operating in 1957. The debt was not cleared for another three years (Fig. 59).

Despite its financial problems, the telescope's public appeal was tremendous. As early as 1953, the dish was becoming a popular attraction, much like those other sites of popular astronomical pilgrimage, Mounts Wilson and Palomar. Asked for more information, the American observatories confirmed that they had a 'serious problem' of 150,000 visitors every year, all needing to be carefully controlled. With growing fears over the potential disturbance to Jodrell Bank's scientists, guards were put on the gate to keep the hordes away. Yet some determined visitors found a way in, including some students who planted a dummy Russian satellite containing a toy dog in the guise of *Sputnik II*'s canine passenger, Laika, addressed to Lovell.

Despite these practical jokes and the financial worries, the dish proved itself as a scientific instrument, helping in the discovery of quasars (which the Hale telescope later identified optically), powerful radiation sources that are extremely distant from Earth and which may provide information about the early history of the universe. So within a short time Jodrell Bank successfully operated both as a powerful research tool and as a beacon for post-war British science.

While Lovell followed his dream of a large dish, radio astronomers at Cambridge pursued a different line of research. Under Martin Ryle, the Cambridge group concentrated on the development of radio interferometry, in which the signals from a number of different receivers are combined to create a single 'image'. Requiring only arrays of simple antennas spread across large areas, Cambridge's instruments avoided the financial and engineering problems of building a single giant dish.

It was in 1967 that the Cambridge radio astronomers achieved a major success during a quasar-hunting project led by Antony Hewish. At this time the main observing array comprised over 1,000 posts connected together by 120 miles (193 km) of wire and cable, spread across almost four and a half acres (1.8 Ha) of land (equivalent to about 57 tennis courts). After just a couple of months of observing, Jocelyn Bell, the researcher in charge of analyzing the 96 ft (29 m) of charts printed out each day, noticed distinct 'scruff', regular groups of pulses coming from the same bit of sky.

At first, their regularity suggested some human origin, perhaps reflections off the Moon or a satellite, or anomalies caused by a corrugated metal building nearby. When the source was established as lying outside the solar system, the astronomers began to wonder if they might be getting signals from an extraterrestrial intelligence. As Bell, who was trying to complete her doctoral thesis, later admitted, she was rather annoyed that 'some silly lot of little green men had to choose my aerial and my frequency to communicate with us'. But as more groups of pulses were detected in other parts of the sky, further analysis showed that they were coming from a new type of astronomical body, a pulsar. These phenomena have since been identified as rapidly rotating neutron stars emitting beams of radio waves. It was for this discovery and the development of the techniques that led to it that Hewish and Ryle were awarded a Nobel Prize in 1974.

Dishes and antennas are still the basic instruments of radio astronomy. The single biggest example is a 300-m (984-ft) dish built into a natural depression at Arecibo in Puerto Rico. Unsurprisingly, these can be costly to build and maintain (Fig. 60). Others capitalize on the fact that the signals from an array of several receivers can be combined to give results equivalent to those from a much larger instrument, greatly reducing construction costs (Fig. 61). By combining the signals from ten dishes located around North America from Hawaii to the Virgin Islands, the Very Long Baseline Array effectively spans more than 5,000 miles (8,000 km), making it the world's largest scientific instrument. Its ability to resolve fine detail has been likened to standing in New York and reading a newspaper in Los Angeles.

The development of radio astronomy and of instruments capable of detecting electromagnetic radiation at other wavelengths has added enormously to astronomers' knowledge of the universe. Whereas observational astronomy in the 1930s was restricted to the wavelengths of visible light (between about 380 and 750 nanometres for a typical person), by the 1960s the range was from 0.001 nanometres to over 10 metres, a huge increase in the 'window' on the universe. In scientific terms, this broader window has had

60 'The Search for Life', by Vince O'Farrell, from *Illawara Mercury*, Australia, 18 January 2004. The biggest telescopes are expensive to build and run. O'Farrell's cartoon raises blunt questions about the world's financial and humanitarian priorities.

major consequences. As Simon Mitton wrote in 1972, the radio astronomy revolution has changed our view of the universe forever:

> the poetic picture of a serene cosmos populated by beautiful wheeling galaxies has been replaced by a catalogue of events of astonishing violence: a primeval fireball, black holes, neutron stars, variable quasars and exploding galaxies.

In creative terms, the development of the radio telescope has given writers and others a spur to imagine new ways in which the universe might be explored in the future. In Victor Appleton II's *Tom Swift and His Megascope Space Prober* (1962), the young crew-cut

Modern Astronomical Telescopes   161

61 The Very Large Array, New Mexico, USA. The observatory has 27 independent radio antennas, each with a 25-m (82-ft) dish. Set roughly in a Y-shape, the dishes combined have an equivalent aperture of 36 km (22.4 miles).

hero builds a revolutionary new telescope that can produce a televisual image of any distant body. This marvellous instrument has an array of technical-sounding components – the 'high-gain amplifier with the helium-extraction device, the anti-inverse-square-wave generator, the wave-terminal equipment, and the radiation "lens"'

– to lend an air of plausibility. It is the sort of techno-speak that the cartoon series *Futurama* mocks and reworks. In 'A Big Piece of Garbage' (1999), Professor Farnsworth invents a 'smelloscope' to explore the universe's odours, but detects a giant ball of malodorous rubbish launched from Earth centuries earlier and which is about to fall back onto New New York. What goes around, comes around; fortunately, our sensing powers improve.

The last sixty years have seen dramatic changes in telescopic astronomy. Astronomers are no longer restricted to collecting visible light, but 'see' the universe in new ways by gathering radio waves, ultraviolet, infrared and X-rays. Having begun as an extension of the human sense of sight, the telescope has broadened the concept of seeing in previously unimagined ways, requiring astronomers no longer to look directly through their instruments, but to analyse data processed by computer interfaces. At the same time, the very form of the instrument has changed. It is hard to look at some of the largest telescopes today and see in them a direct relationship to the small instruments that Galileo and his contemporaries turned to the heavens. The relationship between Galileo's telescope and a radio array is still more obscure.

Epilogue

LOOKING TO THE FUTURE

A telescope will magnify a star a thousand times, but a good press agent can do even better.

Fred Allen, American comedian

In the first years of the seventeenth century, the telescope allowed people to see further than before, across the Earth and out into the universe. Technical advances since then have extended this seeing beyond the limits of human vision. Scientists now probe the cosmos in new ways, revealing a vast, expanding universe. The telescope and its offspring have also enhanced vision in countless more mundane activities, many just as inconceivable four hundred years ago. But where next?

In the world of astronomical telescopes, one thing is certain: the future will be bigger. 'In astronomy, size matters', a leading astronomer recently admitted, 'the question is no longer whether we can build giant telescopes or why we would want to, but when and how large.' The biggest optical telescopes today have mirrors eight to ten metres in diameter; the next generation might be anything from twenty to a staggering one hundred metres across.

Are there limits? Obviously cost, since projects over $1 billion are unlikely to gain support, but also technical problems. Adaptive-optics systems have been a revolutionary step, but it is not clear whether they would work on a 100-m (328-ft) mirror with 100,000 actuators (the small pistons that tune the mirror's shape). Accurate control would be a computing nightmare.

Proposals for radio telescopes are just as ambitious. On completion in 2012, the Atacama Large Millimeter/submillimeter Array (ALMA) will combine the signals from up to 80 antennas 5,000 m (16,400 ft) above sea level in the Chilean Andes, with resolutions ten times better than the Hubble Space Telescope. Looking further ahead, there are plans for a radio telescope with a collecting area of a million square metres, the Square Kilometre Array (SKA). It will link together antenna stations across the world to create an equivalent aperture of several thousand kilometres. While ALMA will provide a bridge between infrared and radio astronomy, SKA will detect low-frequency radio waves, a barely explored region.

The Hubble Space Telescope also has a rich legacy. Plans are well advanced for the James Webb Space Telescope (JWST), a large infrared telescope due to be launched in 2013. Given that the reason to put a telescope in space is to avoid the problems created by Earth's atmosphere, a target further in the future would be to put telescopes on the Moon. The far side of the Moon is protected from terrestrial radio emissions: it's a quiet place to work. A radio telescope built there could also listen in on low frequencies that cannot penetrate Earth's atmosphere.

What will these new telescopes look for? The current Astronomer Royal, Martin Rees, says that the new telescopes will shed light on 'every cosmological era going back to the first light, when the earliest stars (or maybe quasars) condensed out of the expanding debris from the big bang'. It is the big questions that astronomers hope to answer: how the universe originated and evolved; what holds it all together; whether there are other planets that could support life. These are questions that people have asked for generations. Maybe telescopes of the future will give us the answers.

It is less obvious what the future will hold at the smaller end of the scale. Undoubtedly there will be hand-held telescopes of increasing power and sensitivity and with new capabilities. There are already binoculars that can pick out camouflage, and other improvements are certainly being developed for military and security purposes. In time they will make their way into optical devices that everyone can use.

Perhaps all that can be said is that the telescope's story is far from over.

TELESCOPE

There is a moment after you move your eye away
when you forget where you are
because you've been living, it seems,
somewhere else, in the silence of the night sky.

You've stopped being here in the world.
You're in a different place,
a place where human life has no meaning.

You're not a creature in a body.
You exist as the stars exist,
participating in their stillness, their immensity.

Then you're in the world again.
At night, on a cold hill,
taking the telescope apart.

You realize afterward
not that the image is false
but the relation is false.

You see again how far away
each thing is from every other thing.

Louise Glück, from *Averno* (2005)

THE TELESCOPE – A SHORT TIMELINE

1608	Hans Lipperhey and others apply for a patent for a telescope
1609	Galileo Galilei starts making telescopes
	First telescopic observations of the Moon by Thomas Harriot
1610	Galileo Galilei, *Sidereus Nuncius*
1611	The name '*telescopium*' is formally announced
	Johannes Kepler, *Dioptrice*
1637	Copenhagen Observatory founded
1645	Schyrle first describes the terrestrial telescope
1663	James Gregory proposes a design for a reflecting telescope
1667	Paris Observatory founded
1668	Isaac Newton develops a reflecting telescope
1675	Royal Observatory, Greenwich, founded
1721	John Hadley demonstrates a reflecting telescope to the Royal Society
1733	Chester Moor Hall devises the achromatic lens
1758	John Dollond patents the achromatic lens
1781	William Herschel discovers Uranus
1789	Completion of Herschel's 40-ft. reflecting telescope
1819	Cornelius Varley patents the Graphic Telescope
1834	Peter Barlow patents a magnifying lens for use with telescopes
1838	Bessel observes stellar parallax and measures the distance of a star for the first time

1840	Draper photographs the Moon
1845	First observations with the 'Leviathan of Parsonstown'
1846	Discovery of Neptune by Galle and D'Arrest
1854	Porro's patent for a prismatic erecting system
1864	Huggins begins studying stellar spectra
1880	British Royal Navy introduces telescopic sights for heavy guns
1890s	Telephoto lenses applied to cameras
1894	Zeiss begin the mass production of prismatic binoculars
1897	Yerkes Observatory completed
1907	British army officially adopts prismatic binoculars
1917	100-inch Hooker telescope begins observations
1923	Hubble announces that the universe comprises many galaxies
1931–32	Jansky detects radio waves from the sky
1948	Hale telescope dedicated at Mount Palomar
1957	250-foot Jodrell Bank dish begins operating
late 1950s	Quasars observed
1967	Pulsars discovered
1976	Bolshoi Teleskop Azimutal'ny completed
1979	First gravitational lens discovered
1980s	First computerized telescopes for amateur astronomy
1990	Launch of Hubble Space Telescope
1993	First Keck telescope begins observations
1996	Completion of Keck II telescope
1999	Gemini North, Mauna Kea, Hawaii, completed
2001	Very Large Telescope, Cerro Paranal, Chile, completed
2002	Gemini South, Cerro Pachón, Chile, completed
2003	Launch of Spitzer Space Telescope
2005	Large Binocular Telescope, Mt Graham, Arizona, completed

GLOSSARY

aberration – distortion or loss of definition in an image produced by a *lens*; see also *chromatic aberration* and *spherical aberration*
aberration of light – the observed displacement of a star's position that results from the motion of the Earth and the finite velocity of light; also known as astronomical aberration or stellar aberration
achromatic lens – a lens designed to eliminate the effects of *chromatic aberration*
adaptive optics – instrumentation that corrects the blurring of images caused by atmospheric disturbance
altazimuth mounting – a mounting for a telescope that allows it to be moved about a horizontal and a vertical axis
altitude – the angular height of a body above the horizon
aperture – the diameter of the main light-collecting component (the *objective* lens or mirror) of a telescope
astigmatism – a defect of an optical system in which light rays in different planes come to a focus at different points
astrophysics – the branch of astronomy dealing with the physical or chemical properties of celestial bodies
azimuth – an angle measured in the horizontal plane, equivalent to the compass bearing of an object
binocular or **binoculars** – an optical device used with both eyes

camera lucida – an optical drawing aid, patented by William Wollaston in 1806, in which an image is reflected by a prism or mirror supported over a drawing surface, so that the viewer can see both the original image and the drawing

camera obscura – an instrument, sometimes used as a graphic aid, consisting of a darkened chamber or box, into which light is admitted through an aperture or lens, projecting an image onto a surface placed at the focus

charge-coupled device (CCD) – an electronic detector used to measure the *electromagnetic radiation* received by a telescope

chromatic aberration – a defect of lenses, as a result of which different wavelengths of light are brought to a focus at different points, resulting in a coloured fringe around the image produced

coma – an instrumental error in the image of a star that is away from the centre of a telescope's *field of view*, as a result of which the star looks like a comet with a short tail

concave lens – a *lens* that is thinner in the middle than at its edge and which causes a beam of light to diverge

convex lens – a *lens* that is thicker in the middle than at its edge and which causes a beam of light to converge

cosmology – the scientific study of the universe and the laws that govern it

crown glass – glass composed of silica, potash and lime (without lead or iron), which has a low *refractive index*

declination – the angle of a body measured north or south of the celestial equator, equivalent to *latitude* on Earth

double star – two stars that appear sufficiently close to look like a single star

doublet – a *lens* made of two elements

draw tube – any of the tubes that slide into the barrel of a telescope

electromagnetic radiation – radiation in the form of a travelling wave comprising both electric and magnetic fields; includes visible light, infrared waves, ultraviolet light, microwaves, radio waves, gamma rays, and X-rays

equatorial mounting – a telescope mounting in which one axis is directed to the celestial north or south pole

erector or **erecting lens** – a *convex lens* (or group of lenses) placed between the objective and *eye lens* to produce an upright image

eye lens – the *lens* nearest the observer's eye

eyepiece – a *lens* (or group of lenses) placed close to the eye to magnify the image from the *objective* lens

field glass or **glasses** – *binoculars* for use in the field

field lens – an intermediary *lens* used to increase the *field of view* of a telescope

field of view – the angular diameter of the area seen through a telescope or *binoculars*

flea-glass – a small *lens* used for looking at a specimen stuck on a pin

flint glass – glass made of lead oxide, alkali and sand (originally ground flint or pebble), which has a high *refractive index*

focal length – the distance at which light passing through a *lens* (or reflected from a mirror) is brought to a focus

Galilean binoculars – *binoculars* using a Galilean lens system in each tube

interferometer – an instrument for measuring *electromagnetic radiation* by measuring and recording interference patterns

latitude – the angular measurement in degrees of a location north or south of the Earth's equator; used with *longitude* to define a position on the Earth's surface

lens – a piece of glass with one or both sides curved in order to cause light rays to converge or diverge; see also *achromatic, concave, convex, erector, eye* and *objective lens*

light year – the distance travelled by light in one year (approximately 9.5×10^{12} km)

longitude – the distance east or west on the Earth's surface from a *meridian*, measured in degrees, minutes and seconds, or in time (15° being equivalent to one hour)

magnification or **magnifying power** – the number of times larger an object appears when viewed through a telescope or *binoculars* than if viewed with the eye alone; see also *power*

magnitude – a system that ranks stars and other celestial objects according to their brightness. The brightest have low or negative magnitudes. Those of the sixth magnitude are just visible to the naked eye; from the seventh magnitude onwards they are only visible telescopically.

meridian – a circle passing through the Earth's north and south poles and the observer's *zenith*

micrometer – a telescope accessory for measuring small angles

mural quadrant – a *quadrant* mounted onto a wall that is aligned north–south, used to measure the *altitude* of celestial bodies

nanometre – one millionth of a millimetre or 10^{-9} metres

nebula – a celestial body that appears as an indistinct patch of light

neutron star – a very small, dense star comprising mainly neutrons, formed from the collapsed remains of a massive star towards the end of its evolution

nova – a star that becomes more visible due to a dramatic change in its brightness

nutation – the 'nodding' or oscillation of the Earth's polar axis (with a period of 18.6 years)

objective – the *lens* or mirror that is the main light-collecting component of a telescope or *binoculars*

opera glass or **opera glasses** – a small optical instrument with low *magnification*, originally a short telescope and later a small pair of *binoculars*

perspective or **perspective glass** – originally a term for any sort of optical instrument, later mainly applied to short telescopes with a single *draw tube*

piezoelectric crystal – a crystal that can produce an electric potential in response to mechanical stress

power – the capacity of an optical device for increasing the apparent size of an object; the same as *magnification* or *magnifying power*

prismatic binoculars – *binoculars* deploying prisms in the optical system between the *objective* lens and the *eyepiece*

prospective or **prospective glass** – an old term for a telescope or *binoculars*

pulsar – a celestial object that emits regular and rapid pulses of radiation, now recognized as a rapidly rotating *neutron star*

quadrant – an angle-measuring instrument with a graduated arc of 90°

quasar – a 'quasi-stellar object', or celestial object of very small angular size that is a powerful source of electromagnetic radiation

radar – a system for detecting objects by transmitting radio waves and measuring their return after they are reflected

radio telescope – a telescope for detecting and recording radio waves. Radio telescopes may be single concave dishes (which work in a similar way to reflecting telescopes), groups of dishes operating together, or arrays of linked antennas.

refraction – a change in the direction of light (or other) rays as a result of passing between two different media

refractive index – a measure of the *refraction* of a ray of light when it passes between two media

right ascension – an angular measurement of a celestial object's position in the plane of the celestial equator, the equivalent of terrestrial *longitude*

sextant – an instrument with a graduated arc that is equal to a sixth part of a circle, used for measuring angular distances between objects

siderostat – an instrument that compensates for the Earth's rotation, allowing the star that is being observed to remain in the centre of the *field of view*

spectrum – the coloured band into which a beam of light can be split by means of a prism or diffraction grating

spectrograph – an instrument used to record *spectra*

spectroscope – an instrument used to examine *spectra*

spectroscopy – the analysis of *spectra* by instrumental means

speculum – an old term for the mirror of a reflecting telescope

spherical aberration – the blurring of a telescope image formed by a spherical mirror or *lens*, caused by the failure of the light rays to converge at a single point

spiral nebula or **spiral galaxy** – a galaxy consisting of a flat, rotating disk of stars, gas and dust, and a central concentration of stars, the whole surrounded by a much fainter halo of stars

spyglass – an alternative name for a terrestrial telescope, often referring to a small Galilean telescope

stellar parallax – a change in the observed position of a star caused by a change in the observer's position

supernova – a star that is exploding at the end of its evolution

surveyor's level – an instrument used in surveying to measure the height of distant points in relation to a bench mark

theodolite – a surveying instrument, originally for measuring horizontal angles, now also used for measuring vertical angles

transit instrument – any astronomical instrument in which the positions of celestial objects are measured by observing them as they cross the observer's *meridian*; includes transit telescopes and transit circles

triplet – a *lens* made of three elements

variable star – a star whose brightness varies, either periodically or irregularly

zenith – the point in the sky directly above the observer

zenith sector – a telescopic instrument that points directly overhead, allowing the angular distance of a star from the observer's *zenith* to be measured

Refracting telescopes (or **refractors**) use lenses to gather and focus light.

Galilean telescope – named after Galileo Galilei (1564–1642). The lens system has a convex objective lens and a concave eye lens, a combination that produces an erect image but has a limited field of view. The first telescopes were of this type. Many cheap telescopes still use this system.

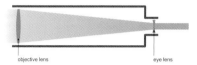

Keplerian or **astronomical telescope** – named after Johannes Kepler (1571–1630), who described the system in the *Dioptrice* (1611). The system uses two convex lenses and has the advantage of a wider field of view than the Galilean telescope, but produces an inverted image.

Terrestrial telescope – first mentioned by Anton Maria Schyrle de Rheita (1597–1660), with two lenses added to the Keplerian arrangement to produce a telescope with a good field of view and an erect image, which was more practical for terrestrial purposes. By the end of the eighteenth century, terrestrial telescopes had four or more lenses in the eyepiece assembly, with four becoming standard.

Reflecting telescopes (or **reflectors**) use a mirror as the main light-gathering element, with additional mirrors and/or lenses to produce the final image. The main systems are:

Gregorian telescope – named after James Gregory (1638–75), who described the arrangement in 1663. The system uses a small concave mirror to reflect the light back through a hole in the primary mirror, where it can be viewed with an eyepiece.

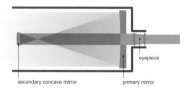

Newtonian telescope – the system used by Isaac Newton (1643–1727) in 1668. The image from the primary mirror is reflected by a small angled mirror and viewed from the side of the instrument.

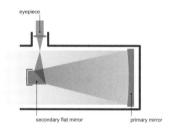

Cassegrainian telescope – described by Laurent Cassegrain (1629–93) in 1672, the system uses a small convex mirror to bring the light from the primary mirror to a focus. Like the Gregorian system, the image is viewed through a hole in the primary mirror.

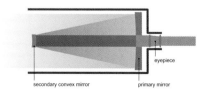

FURTHER READING

Jon Agar, *Science & Spectacle. The Work of Jodrell Bank in Post-War British Culture* (Amsterdam, 1998)
Geoff Andersen, *The Telescope. Its History, Technology, and Future* (Princeton & Oxford, 2007)
H. Barty-King, *Eyes Right. The Story of Dollond & Aitchison 1750–1985* (London, 1986)
J.A. Bennett, *The Divided Circle* (Oxford, 1987)
Mario Biagioli, *Galileo, Courtier* (Chicago, 1993)
Mario Biagioli, *Galileo's Instruments of Credit* (Chicago & London, 2006)
Reginald Cheetham, *Old Telescopes* (Southport, 1997)
T.N. Clarke, A.D. Morrison-Low & A.D.C. Simpson, *Brass & Glass. Scientific Instrument Making Workshops in Scotland as Illustrated by Instruments from the Arthur Frank Collection at the Royal Museum of Scotland* (Edinburgh, 1989)
T.H. Court & M. von Rohr, 'A History of the Telescope from about 1675 to 1830 Based on Documents in the Court Collection', *Transactions of the Optical Society*, 30 (1929), pp. 207–260
Michael J. Crowe, *The Extraterrestrial Life Debate 1750–1900* (Cambridge, 1986)
Terence Dickinson & Alan Dyer, *The Backyard Astronomer's Guide* (Willowdale, 2002)

Stillman Drake, *Discoveries and Opinions of Galileo* (New York, 1957)
Benjamin Elman, *A Cultural History of Modern Science in China* (Cambridge, Mass. & London, 2006)
Patricia Fara, *Pandora's Breeches: Women, Science and Power in the Enlightenment* (Pimlico, 2004)
Eric Forbes, A.J. Meadows & Derek Howse, *Greenwich Observatory* (London, 1975)
Christopher Frayling, *Mad, Bad and Dangerous? The Scientist and the Cinema* (London, 2005)
Carl C. Gaither & Alma E. Cavazos-Gaither, *Astronomically Speaking. A Dictionary of Quotations on Astronomy and Physics* (Bristol, 2003)
Galileo Galilei, *Sidereus Nuncius or the Sidereal Messenger* (tr. Albert van Helden) (Chicago, 1989)
Owen Gingerich (ed.), *The General History of Astronomy. Volume 4 Astrophysics and Twentieth-century Astronomy to 1950: Part A* (Cambridge, 1984)
D. Gooding *et al.* (eds), *The Uses of Experiment* (Cambridge, 1989)
John H. Hammond, *The Camera Obscura: A Chronicle* (Bristol, 1981)
Roslynn D. Haynes, *From Faust to Strangelove. Representations of the Scientist in Western Literature* (Baltimore & London, 1994)
Richard Holmes, *The Age of Wonder* (London, 2008)
Holly Henry, *Virginia Woolf and the Discourse of Science* (Cambridge, 2003)
Richard Holmes, *The Age of Wonder* (London, 2008)
Michael Hoskin (ed.), *The Cambridge Illustrated History of Astronomy* (Cambridge, 1997)
Michael Hoskin, *The History of Astronomy. A Very Short Introduction* (Oxford, 2003)
Michael Hoskin, *The Herschel Partnership* (Cambridge, 2003)
Myles Jackson, *Spectrum of Belief. Joseph von Fraunhofer and the Craft of Precision Optics* (Cambridge, Mass. & London, 2000)
Lisa Jardine, *On a Grander Scale. The Outstanding Life of Sir Christopher Wren* (London, 2002)
Martin Kemp, *The Science of Art. Optical Themes in Western Art from Brunelleschi to Seurat* (New Haven, 1990)
Henry King, *The History of the Telescope* (London, 1955)
Rudolf Kingslake, *A History of the Photographic Lens* (Academic Press, 1989)
Charlotte Klonk, *Science and the Perception of Nature* (New Haven, 1996)
John Lankford (ed.), *History of Astronomy. An Encyclopedia* (New York & London, 1997)
David Levy, *David Levy's Guide to the Night Sky* (Cambridge, 2001)
Timothy Lenoir (ed.), *Inscribing Science* (Stanford, 1998)

Bernard Lightman (ed.), *Victorian Science in Context* (Chicago, 1997)

Anita McConnell, *Jesse Ramsden (1735–1800) London's Leading Scientific Instrument Maker* (Aldershot, 2007)

W. Patrick McCray, *Giant Telescopes. Astronomical Ambition and the Promise of Technology* (Cambridge, Mass. & London, 2004)

W. Patrick McCray, *Keep Watching the Skies! The Story of Operation Moonwatch and the Dawn of the Space Age* (Princeton, 2008)

A.J. Meadows, *The High Firmament. A Survey of Astronomy in English Literature* (Leicester, 1969)

John Millburn, *Retailer of the Sciences* (London, 1986)

Patrick Moore, *Eyes on the Universe. The Story of the Telescope* (London, 1997)

Alison Morrison-Low, *Making Scientific Instruments in the Industrial Revolution* (Aldershot, 2007)

Stephen Moss, *A Bird in the Bush. A Social History of Birdwatching* (London, 2004)

J. Needham & C. Ronan, *The Shorter Science and Civilisation in China*, vol. 2 (Cambridge, 1981)

Marjorie Nicholson, *Science and Imagination* (New York, 1956)

Roberta J.M. Olson & Jay M. Pasachoff, *Fire in the Sky* (Cambridge, 1998)

Richard Panek, *Seeing and Believing* (London, 2000)

Patricia Phillips, *The Scientific Lady* (London, 1990)

Colin Pillinger, *Space is a Funny Place* (London, 2007)

F.K. Pizor & T.A. Comp, *The Man in the Moone. An Anthology of Antique Science Fiction and Fantasy* (London, 1971)

Richard Preston, *First Light. The Search for the Edge of the Universe* (London, 1992)

Eileen Reeves, *Painting the Heavens. Art and Science in the Age of Galileo* (Princeton, 1997)

Eileen Reeves, *Galileo's Glassworks: The Telescope and the Mirror* (Cambridge, Mass. & London, 2008)

Rolf Riekher, *Telescopes and their Masters* (forthcoming)

Adam Roberts, *The History of Science Fiction* (Basingstoke, 2007)

Edward Rosen, *The Naming of the Telescope* (New York, 1947)

Edward Rosen (tr. and ed.), *Kepler's Conversation with Galileo's Sidereal Messenger* (London & New York, 1965)

Simon Schaffer, 'Instruments, Surveys and Maritime Empire', in D. Cannadine (ed.), *Empire, the Sea and Global History* (Basingstoke, 2007), pp. 83–104

Timon Screech, *The Western Scientific Gaze and Popular Imagery in Later Edo Japan* (Cambridge, 1996)

William Sheehan, *Planets and Perception. Telescopic Views and Interpretations, 1609–1909* (Tucson, 1988)

J. Shirley (ed.), *Thomas Harriot Renaissance Scientist* (Oxford, 1974)

Robert W. Smith, *The Space Telescope* (Cambridge, 1989)

Albert van Helden, 'The Invention of the Telescope', *Transactions of the American Philosophical Society*, 67 (1977)

Deborah Jean Warner, 'Telescopes for land and sea', *Rittenhouse*, 12 (1998), pp. 33–54

Fred Watson, *Binoculars, Opera Glasses and Field Glasses* (Princes Risborough, 1995)

Fred Watson, *Stargazer. The Life and Times of the Telescope* (Crows Nest & London, 2004)

J.B. Zirker, *An Acre of Glass. A History and Forecast of the Telescope* (Baltimore, 2005)

Periodicals
Annals of Science
Bulletin of the Scientific Instrument Society
Burlington Magazine
History of Science
History Today
Isis
Journal for the History of Astronomy
Journal of the Antique Telescope Society
Journal of the History of Ideas
Notes and Records of the Royal Society of London
The Observatory
Osiris
Oxford Dictionary of National Biography
Quarterly Journal of the Royal Astronomical Society
Science in Context
Sky & Telescope
Vistas in Astronomy

ACKNOWLEDGEMENTS

Many people helped in the preparation of this book. Among those, I am very grateful to everyone who gave advice and pointed out interesting sources, including Alexi Baker, Graham Dolan, Mike Dryland, Ruth Dunn (and her many assistants), Brian Gee, Hans Hooijmaijers, Melanie Keene, Karen Lund, Don Pugh, Nicky Reeves, Gilbert Satterthwaite, Simon Schaffer, Janet Small, Pat Wainwright, Emily Winterburn and the subscribers to *Rete*. I would also particularly like to thank everyone who commented on earlier drafts, notably Peter Abrahams, Marv Bolt, Ron Bristow, Gloria Clifton, Rebekah Higgitt, Marek Kukula, Meryl Pugh and Nigel Rigby, although all errors remain my own. Lastly, particular thanks are due to Emily Winter and Sara Ayad at NMM Publications, and to Tina Warner, Ben Gilbert, and Brendan Hennessy from NMM Photographic Services.

Many of the texts consulted while researching this book are listed in the 'Further Reading', although the list is not exhaustive. Among the many academic papers not individually mentioned there, I wish to highlight the work of Jim Bennett and Albert van Helden, both of whom have published extensively on many aspects of the telescope's history.

PICTURE CREDITS

All illustrations, except where stated, are © National Maritime Museum, Greenwich, London, and may not be reproduced without permission. They may be ordered from the NMM Picture Library, National Maritime Museum, Greenwich, London SE10 9NF, quoting the image reference number given after the page number below. (See www.nmmimages.co.uk for full contact details.)

Figs: 1 F8652; 2 F8596; 3 F8665; 4 PT3458; 5 E5556-4; 6 D7124-8; 7 BHC 2700; 8 F8630; 9 F8641-4; 10 F8624; 11 E5556-4; 12 F8619; 13 F8619; 14 BHC1812; 16 BHC3025; 19 F2985; 20 PW3526; 21 PY6061; 22 F5601-1 (and titlepage); 23 F8607; 24 F8629; 25 F8634-1,2; 26 PT3274; 27 F4451-1; 28 F4487-1; 29 F8672; 30 PW3997; 31 PW3751; 32 F4520-1; 33 PW2940; 34 F5926; 35 F8661; 36 B5698-C; 37 B1646; 38 B1646; 39 PU1906; 41 PW5136; 44 PWF3824; 45 F8775; 46 F8649; 48 PW3726; 49 F8737-1.

We would also like to thank the following for permission to include their images: Bayerische Staatsgemäldesammlungen/Alte Pinakothek, München © BPK, *Fig.* 15

BFI Archives, London, by kind permission, *Fig.* 50
© Karen Lund, *Fig.* 51
Museo del Prado, Madrid/Joseph Martin/© akg-Images, London, *Fig.* 17
© NASA, *Figs* 57; 58
National Radio Astronomy Observatory, New Mexico © NRAO/AUI, *Fig.* 61
Pennsylvania State University, PA, courtesy Rare Books and Manuscripts Division, © 2008, *Fig.* 18
politicalcartooons.com © Copyright 2004 O'Farrell – All Rights Reserved, *Fig.* 60
© PRM Wainwright, *Fig.* 53
© Punch/David Langdon, *Fig.* 59
© Richard Dunn, *Figs* 40; 42; 43; 47; 52; 54; 55
© Richard Wainscoat, *Fig.* 56

Diagrams by Greg Smye-Rumsby, Concept Design, *pages* 175–76

INDEX

Page numbers in *italics* refer to captions to the illustrations.

aberration 28, 80, 89
– chromatic 57, *75*, 76–8
– of light 89, *90*
– spherical 77, 150
Accademia dei Lyncei 31
achromatic lens 77–8, 80, *82*, 84, 88
adaptive optics 151–2, 165
Admiralty, *see also* Royal Navy 86, 102, 130
Adriaenszoon, Jacob, *see* Metius, Jacob
aerial telescope, *see* telescopes, types of
Airy, George Biddell (1801–1892), seventh Astronomer Royal 93, *94*, 99
Albumazar (play) 52
Alcock, George (1912–2000) 139
Alice's Adventures in Wonderland (book) 109, 119
Almagest (book) 16
Almagestum novum (book) 6
amateur astronomy 136, 138–9, 150
American Association of Variable Star Observers 138

Anatomy of the World, An (poem) 48
Anderson, John (1876–1959) 146
Andromeda 143, 148
Archimedes (*c.*287–*c.*212 BC) 12
Aristophanes (*c.*486–*c.*386 BC) 12
Aristotle (384–322 BC) 12, 15–19
As Seen Through the Telescope (film) *133*
astrolabe *13*, 16, *17*
astrology 15, 50
Astronomical Society of London, *see also* Royal Astronomical Society 95
astrophotography 149
astrophysics 97, 99, 141–2
Aubert, Alexander (1730–1805) 65
Australia 104–5, 131, 138, 156, *161*

Baade, Walter (1893–1960) 148, 150
Bacon, Roger (*c.*1214–1292?) 11–14
Barlow, Peter (1776–1862) 132
Bass, George 78
Bath 60, 66

Bayly, William (1738–1810) *101*, 102
Behn, Aphra (1640?–1689) 52
Bell, Jocelyn (1943–) 159–60
Bell, Johan Adam Schall von *see* Schall von Bell, Johann Adam
Bell Telephone Laboratories 156
Bellarmine, Cardinal (1542–1621) 30
Benoist, Michel (1715–1774) 105
Bessel, Friedrich (1784–1846) 92
Biggs, James 154
binoculars 127–8, *129*, 130–6, *137*, 138–9, 166
– binocular opera glasses 128
– binocular telescopes 85
– Galilean 129
– prismatic 129, *131*, 135
– trench 130
Bird, John (1709–1776) 73, 89
Birds through an Opera Glass (book) 135
bird-watching 136
Birr Castle 95
Blackett, Patrick (1897–1974) 157
Boer War 129
Boland, Edward P. (1911–2001) 154
Bologna, University of 30, 56
Bond, George Phillips (1825–1865) 97
Boyle, Robert (1627–1691) 53
Bradley, James (1692–1762), third Astronomer Royal 58, 89, *90*, 92
Brahe, Tycho (1546–1601) 18, 38–9
Braine, Robert 122, 124
Brevissima peregrinatio contra nuncium sidereum (book) 29
Brisbane, James 104–5
Brisbane, Thomas (1773–1860) 104
Brownstein, Gabriel 134
Brueghel 'the Elder', Jan (1568–1625) 46, *48*
Bungaree (Bongaree) 105
Burney, Frances [Fanny] (1752–1840) 68
Butler, Samuel (1613–1680) 53

Caesar, Julius (100–44 BC) 14
California Institute of Technology 147
Cambridge University 156, 159
camera 118, 132
camera lucida 110
camera obscura 110
Campani, Giuseppe (1635–1715) 37
Caravaggi, Cesare 56
Carnegie, Andrew (1835–1919) 143
Carnegie Institution 142
Carroll, Lewis (Charles Dodgson, 1832–1898) 109, 119
Carte du Ciel 98
Cassini, Giovanni (1625–1713) 43
Cassini, Jean Dominique de (1748–1845) 67
Cavalieri, Bonaventura (*c.*1598–1647) 56
Cavendish, Margaret, Duchess of Newcastle (1623?–1673) 53–4
charge-coupled devices (CCDs) 150
Christie, William (1845–1922), eighth Astronomer Royal 99
Christmann, Jacob (1554–1613) 29
Cigoli, Ludovico (1559–1613) 45
Clavius, Christopher (1537–1612) 30
Clerke, Agnes Mary (1842–1907) 118
Clouds, The (play) 12
Cold War 152, 157
Columbus, Christopher (1451–1506) 47, 124
comets 68, 138, 150
Compleat System of Opticks (book) 61
Conversation, The (film) 133
Cook, Captain James (1728–1779) *101*, 102
Copenhagen 42
Copenhagen, Battle of
– 1801 *112*, 113
– 1807 *84*
Copernicus, Nicolaus (1473–1543) 17–19
Coppola, Francis Ford (1939–) 133
Cotman, John Sell (1782–1842) 110
Crete 12
Crimean War 128
Curious Case of Benjamin Button, Apt. 3W, The (book) 134

Danzig (Gdańsk) 37–8, 40
Dee, John (1527–1609) 15
Demisiani, John 32
De revolutionibus orbium coelestium (book) 17
Description of a New World, Called the Blazing World, The (book) 53
Diaz, Emmanuel 33
Dickens, Charles (1812–1870) 105
Digges, Leonard (*c.*1515–*c.*1559) 14, 23
Digges, Thomas (*c.*1546–1595) 14, 23
Dioptrice (book) 21, 34
Discovery of the World in the Moon, The (book) 49
Dissertatio cum Nuncio Sidereo (book) 49–50
Divini, Eustachio (1610–1685) 37
Dollond, optical instrument makers 72–3, 75–9, 80, *81*, 82, 84, 86–7, 105
Dollond, George (1774–1852) 79
Dollond, John (1707–1761) 74, *75*, 76–7, *78*
Dollond, Peter (1731–1820) 74, 77, 79, 80, *82*
Dollond & Aitchison 82
Dollond & Co. 87
Donne, John (1572–1631) 48
double stars 64, *100*
Draper, John (1811–1882) 97

East India Company 33, 102, 105
Economy of the Eyes (book) *10*, 85
Egypt 12–13, 45–6
Elephant in the Moon, The (poem) 53
Elsheimer, Adam (1578–1610) 45, *46*
Emblemes (book) 51
Emperor of the Moon, The (play) 52
Enemy of the State (film) 132
Entretiens sur la Pluralité des Mondes (book) 116
Euclid (fl. 300 BC) 12
Euler, Leonhard (1707–1783) 76
European Space Agency 153
Evans, Robert (1937–) 138–9
Evelyn, John (1620–1706) 53
Experimental Philosophy (book) 50

Exposition Universelle, Paris (1900) 106, *107*
eyeglasses 14

field glasses 128, 130
First World War 129–30, 150
Flamsteed, John (1646–1719), first Astronomer Royal 38, 40, *41*, 43, 67, 92, *138*
Flinders, Matthew (1774–1814) 105
Florence 14, 24, 26, 135
Fontenelle, Bernard le Bovier de (1657–1757) 116
Foucault, Léon (1819–1868) 141
Fraunhofer, Joseph (1787–1826) 91–2, 98
Futurama (TV series) 163

Galilean telescope, *see* telescopes, types of
Galilei, Galileo (1564–1642) 10, 22, *24*, 25–6, *27*, 28–33, 37, 45, *46*, 47–50, 52, 56, 64, 67, 78, 124, 130, 163
Gascoigne, William (1612?–1644) 38
George III (1738–1820) 64–5, *84*, *113*, 119
glass-making 13, 91
Godwin, Francis (1562–1633) 49
GPS (Global Positioning System) 132
Graham, George (*c.*1673–1751) 73, 88–9
Graphic Telescope, *see* telescopes, types of
Great Exhibition (1851) 106
Greeks, ancient 12–16, 64
Green, Nathaniel (1823–1899) 124
Greenwich *see also* Royal Observatory 39–40, *41–2*, 43, 63, 65–6, 71, 80, *82*, 88–9, *90*, 92–3, *94*, 99, *100*, 104, 125
– Meridian 93
– Mean Time 93, *94*
Gregory, David (1659–1708) 76
Gregory, James (1638–1675) 56–7
Gresham College, London 40
Grosseteste, Robert (*c.*1170–1253) 13
Guinand, Henri (1771–1852) 92
Guinand, Pierre Louis (1748–1824) 91–2
Gulliver's Travels (book) 118–19
Gustavus Adolphus (1594–1632) 36

Hadley, John (1682–1744) 57–8, 61
Hale, George Ellery (1868–1938) 141–7, 152–4
Hale telescope, *see* telescopes, specific
Hall, Chester Moor (1703–1771) 78
Halley, Edmond (1656–1742), second Astronomer Royal 38, 88–9
Hardy, Thomas (1840–1928) 121–2
Harriot, Thomas (*c.*1560–1621) 26, 28
Herschel, Alexander (1745–1821) 62, 66
Herschel, Caroline (1750–1848) *60*, 61–3, 65, 67–8, 70–1
Herschel, John (1792–1871) 70, 102, *104*, 123
Herschel, William (1738–1822) 59–71, 88, 93, *104*, 121, 124, 157
Herschel telescope *see* telescopes, specific
Hevelius, Elisabeth (1647–1693) 37
Hevelius, Johannes (1611–1687) 37–8, 40, 42
Hewish, Antony (1924–) 159–60
Hitchcock, Alfred (1899–1980) 127, 133
Hoffmann, E.T.A. (1776–1822) 119, 121
Hogarth, William (1697–1764) *8*
Holmes, Oliver Wendell (1809–1894) 68
Hooke, Robert (1635–1703) 38, 40, 42, 49–50, 53, 57
Hooker, John D. (1837–1910) 143
Hooker telescope *see* telescopes, specific
Horky, Martin 28–30
Hornor, Thomas (1785–1844) 110–11
Huahine (Polynesian island) 102
Hubble, Edwin (1889–1953) 143–5, 148, 152
Hubble Space Telescope *see* telescopes, specific
Hubble Ultra Deep Field *155*
Huggins, William (1824–1910) 98–9
Humours of Oxford, The (play) 117–18
Huygens, Christiaan (1629–1695) 29, 35, 37–8, *39*, 40, 58
Huygens, Constantijn (1628–1697) 58

Ignatius His Conclave (book) 48
India 102
infrared 155, 163, 165–6

Inquisitive Giant, The (film) 157
instrument making 75–6, 79
International Meridian Conference (1884) 93
Islamic philosophy 13

Jansky, Karl (1905–1950) 156
Janssen, Sacharias (*c.*1585–*c.*1632) 22
Japan 33, 113, 131
Jeans, James (1877–1946) 143–4
Jellicoe, Admiral John (1859–1935) 130
Jonson, Ben (1572–1637) 44, 49
Jupiter 26, 58, 155
– moons 26, 38, 30, 34, 49

Kepler, Johannes (1571–1630) 20, 27–8, 30, 34, *35*, 49–50
Kirchoff, Gustav (1824–1887) 98
Kitchiner, William (1778–1827) *10*, 85–6
Klingenstierna, Samuel (1698–1765) 76

Lagalla, Giulio Cesare (1576–1624) 30
Laika 159
Lassell, William (1799–1880) 96
latitude 101
lenses
– achromatic 77–8, 80, *82*, 84, 88
– concave 14, 32
– convex 14, 32, 34–5
– development 37, 78
– magnifying 132
– manufacture *10*, 14, 37, 91
– objective 93, 99, *39*, 86
– quality 14, 23, 29
– telephoto 132, 134
– triplet 80
– zoom 132
lens-grinders *10*, 35, 37, 78, 150
Leviathan of Parsonstown *see* telescopes, specific
Levy, David (1948–) 139
Li Yu (1610–1680) 34
Lichtenberg, Georg Christoph (1742–1799) 66

Index 187

Lipperhey, Hans (d. 1619) 22–6, 56, 128, 130
Locke, Richard Adams (1800–1871) 123
London
– craft workers 57
– Great Fire 40
– lens-makers 56
– opticians 77–9, 82
– scientific instrument makers 73, 79, 88–9, 91
longitude 43, 101
Longland, William (d. c.1722) 36
Louis XIV (1638–1715) 37
Lovell, Bernard (1913–) 156–9
Lowell, Percival (1855–1916) 124–5
Lower, Sir William (c.1570–1615) 47

Macartney, Lord George (1737–1806) 105–6
Macmillan, Harold (1894–1986) 157
Magini, Giovanni Antonio (1555–1617) 30
Manchester University 157
Man in the Moone, The (book) 49
Mann, James (1683/4–1757) 78
Mars
– canals of 124, *125*
– orbit 20
Martin, Benjamin (1705–1782) 77, *117*, 118–9
Martineau, Harriet (1802–1876) 95
Maruyama Ōkyo (1733–1795) 112
Maskelyne, Nevil (1732–1811), fifth Astronomer Royal 65–6, 80, 88–90
Meade Instruments Corporation 136
Mercury 64
Merriam, Florence (1863–1948) 135
Mersenne, Marin (1588–1648) 56
Merz 93
Messages from Mars, By the Aid of the Telescope Plant (book) 122
Metius, Jacob (1571–1628) 22, 50
Micrographia (book) 49
micrometer 38
microscope 49–50, 126
Milky Way 26, 45, *46*, 143, 156

Miller, James (1704–1744) 117
Miller, William 98
Milton, John (1608–1674) 48
Miriam's Schooling (book) 122
mirrors 12, 15, 20, 29, 56, 58–9, 63, 66, 97, 142, *149*, 151
– manufacture 14, 148
– telescope 87, 57, 61, 94–5, 105, 110, 141, 165
Mitton, Simon 161
Molyneux, William (1656–1698) 38
Moon 8, 19, 26, 28, 31, 37, 45, *46*, 48–50, 53, 93, 97, 101, 108, 123–4, 136, 159, 166
– hoax (1835) 123
Morrison, Lieutenant R.J. (1795–1874) 20
mural quadrant 88–9
Mysterious Universe, The (book) 143

NASA 153–4
Nasmyth, James (1808–1890) 96
Nassau, Maurice of (1567–1625) 22–3
nebulae 64, 67–8, 95, 97–8
Nelson, Horatio (1758–1805) *112*, 113–14
Neptune 96
Newbery, John (1713–1767) 116
News from the New World Discovered in the Moon (play) 44, 49
Newton, Sir Isaac (1642–1727) 57, 61, 76, 114
Newtonian system of philosophy, The (book) 116
New York University 97
Nicholson, Max (1904–2003) 135
Nile, Battle of the 113
Nimrud 12

observatories
– Arecibo 160
– Brera 124
– Cape 102, *104*, 123
– Chandra X-ray Observatory 155
– Copenhagen 42
– Dorpat 92

- European Southern Observatory 148
- Flagstaff 124
- Göttingen 66
- Hamburg 150
- Harvard College 97
- Hven 19
- Jodrell Bank 156–7, *158*, 159
- Kew 65, 86
- Kitt Peak 148
- Königsberg 92
- Lick 99
- Madras 102
- Madrid 66
- Mount Palomar 146, 148, 150
- Mount Wilson 142–4, *145*, 146, 150
- Palermo 66, 89, 91
- Parramatta 104
- Paris 43, 67, 76
- Pulkovo 99
- Royal Observatory, Greenwich 40, *41*–2, 63, 65, 71, *82*, 88, *90*, 92, *94*, 99, *100*, 125
- St Petersburg 66
- US Naval Observatory 99
- Williams Bay 99
- Yerkes 99, 141–2

Oculus Enoch et Eliae (book) 34
Optica Promota (book) 13, 56
optics 46, 57, 74, 76, 91, 132, 151
- adaptive 151–2, 165
Optics (book) 12
opticians 77–9, 82, 92, 150
Optisches Institut (Optical Institute) 91
Opus Maius (book) 11, 14

Paradise Lost (poem) 48–9
Parsons, William (1800–1867) 95–6, 98–9
Pepys, Samuel (1633–1703) 36, 56
Pharos of Alexandria 14
photography 97–9, 150
Pisa, University of 24, 27
Pitt, William (1759–1806) 66
Place, Francis (1647–1728) *41*
Playfair, Lyon (1818–1898) 106

Pliny the Elder (23–79 AD) 13
Pond, John (1767–1836), sixth Astronomer Royal 66, 88, 90
Porro, Ignazio (1801–1875) 128
Pound, James (1669–1724) 58
Princeton University 153
prisms 75, 77, 128–9
Prolusiones (poem) 51
Ptolemy (after 83–*c.*168 AD) 16
pulsars 160

Qianlong Emperor (1711–1799) 105
Quarles, Francis (1592–1644) *51*
quasars 159, 161, 166

radar 156–8
radio arrays 163
- Atacama Large Millimeter/submillimeter Array (ALMA) 165
- Square Kilometre Array (SKA) 165
- Very Large Array *162*
- Very Long Baseline Array 160
radio astronomy 156–7, 160–1, 165
radio waves 160, 163, 165
Ramsden, Jesse (1735–1800) 79, 89, 91, 105
rangefinder 130–1
Rear Window (film) 127, 133–4
Reber, Grote (1911–2002) 156
Rebeschke, Katharine 37
Rees, Martin (1942–) fifteenth Astronomer Royal 166
Reeve, Richard (d. 1666) 56
reflecting telescope, *see* telescopes, types of
refracting telescope, *see* telescopes, types of
Research Enterprises Ltd *131*
Rest on the Flight into Egypt, The (painting) 45, *46*
Rheita, Anton Maria Schyrle de (1604–1659/60) 35
Riccioli, Giovanni Battista (1598–1671) *6*
Ritchey, George (1864–1945) 142
Rockefeller Foundation 146–7
Romans, ancient 13

Rome 30–1, 45
Rome, University of 30
Rømer, Ole (1644–1710) 89
Rowlandson, Thomas (1757–1827) *115*, 119, *120*
Royal Astronomical Society 68, 70, 95
Royal Hospital School, Greenwich 125
Royal Navy, *see also* Admiralty 80, 104, 129, 130
Royal Society 38, 42, 53, 57–8, 76–7, 95, 116, 118
Rubens, Peter Paul (1577–1640) 46, *48*
Rudolf II, Emperor (1552–1612) 20
Ryle, Martin (1918–1984), twelfth Astronomer Royal 159–60

Sagredo, Giovanfrancesco (1571–1620) 29, 33
Sandmann, Der (short story) 119
Saris, John (c.1580–1643) 33
Sarpi, Paolo (1552–1623) 25
satellites
– reconnaissance 132, 151–2, 158
– Sputnik 157, 159
Saturn 70, 142
– moons 29, 58
– rings 29
Scarlett, Edward (c.1703–c.1779) 78
Schall von Bell, Johann Adam (1591–1666) 34
Scheiner, Christoph (1573–1650) 31, 34, *35*
Schiaparelli, Giovanni (1835–1910) 124
Schmidt, Bernhard (1879–1935) 149–50
Schorr, Richard (1867–1951) 150
Schyrle, Johann Burchard, *see also* Rheita, Anton Maria Schyrle de 35
science fiction 122
Scott, Ridley (1937–) 132–3
Searchlight, The (short story) 144
Secchi, Father Angelo (1818–1878) 99
Second World War 130, 132, 136, 143, 153, 156
Selenographia (book)

Seneca (c.4 BC–AD 65) 13
sextant 38, 100, 102
Shadwell, Thomas (c.1640–1692) 53
Short, James (1710–1768) 58, *59*, 67, 77
Shovell, Sir Cloudesley (1650–1707) *47*
Sidereus Nuncius (book) 26, *27*, 31, 45, 52
Sight (painting) 46, *48*
sighting tube *13*, 89
Sirturus, Hieronymous 33
Sky & Telescope (magazine) 136
Sky at Night, The (TV show) 136
Smith, Robert (1689–1729) 61
spectacles 14, 73
spectroscopy 93, 97–9
Spitalfields Mathematical Society 76
Spitzer, Lyman Jr (1914–1997) 153, 155
spyglass, -es 25–6, *78*, *84*, *113*, 119
St Paul's Cathedral, London 110
Stapledon, Olaf (1886–1950) 144–5
Star Maker (book) 143–4
stellar parallax 19, 40, 92
Struve, Friedrich Georg Wilhelm (1793–1864) 92
Sun 8, 17, 19–20, 31, *35*, 89, 92, 101, 156
– sunspots 31
supernovae 138–9
surveillance 10, 132–3, 139
surveying 85, 100, 105, 108, 132
Syracuse 12

telescope, alternative names for
– *Batavica dioptra* 32
– *instrumentum* 32
– *occhiale* 32
– opera glass 36, 85–6, 128
– optic engine 53
– *organum* 32
– perspective 36, 45, 86
– perspective glass 15, 45
– *perspicillum* 32
– prospect glass 33, 45, 85
– spyglass 24–6, 28, 33, 73, *78*, *84*, 86, *113*, 119

– *telescopium* 32
– trunk 26, 44, 49, 53
telescopes, specific
– Airy's transit circle 94
– Bolshoi Teleskop Azimutal'ny 148
– Flamsteed's well telescope *41*
– Grande Lunette 106, *107*, 108
– Hale *147*, 148, 150, 159
– Herschel's 40-foot *68*, 97
– Herschel's large 20-foot 63, *104*, 121
– Herschel's small 20-foot 63–4
– Hooker 143, 146
– Hubble Space Telescope *152*, 155, 165
– James Webb Space Telescope (JWST) 165
– Keck *149*
– Leviathan of Parsonstown 84, *96*, 97
– Mayall 148
– Monument 40
– Oschin Schmidt 150
– Spitzer Space Telescope 155
telescopes, types of
– aerial *39*, 58
– Cassegrain 152
– Dutch, *see* Galilean
– Galilean 28, *32*, 34, 56
– Go To 137
– Graphic 110, *111*
– Keplerian (or astronomical) 34–5
– marine 86, 91
– Night or Day 86
– Officer of the Watch 86, *87*
– reflecting 56–8, *59*, 61, *62*, 63–6, 71, 88, 93, 95, 101, *104*, 105, 138, 141–2, 148, *152*
– refracting 21–43, 58, 61, 63, 73, 76–7, 88, 92, 99, 105–8, 141–2
– Schmidt 150
– space *152*, 153–5, 165
– terrestrial 35–7
– transit 80, *82*, 89, 90, 102, 104
– well *41*
– zenith 89, *90*, 105

telescopic sights 38–9, 130, 132
Tom Swift and His Megascope Space Prober (book) 161
'Tom Telescope' 116
Tomkis, Thomas (*c.*1580–*c.*1615) 52
Torricelli, Evangelista (1608–1647) 37
Trafalgar, Battle of 80, *112*
Troughton & Simms 93, *102*
Troy 12
Two on a Tower (book) 121

ultraviolet 153, 163
Umbroc 131
Uranus 59, 64–5, 105
US Congress 154
Utzschneider, Joseph von (1763–1840) 91

Varley, Cornelius (1781–1873) 110, *111*
Varley, John (1778–1842) 110
Venice 14, 33
Venus
– phases 31, 34, 52, 67
– transit of 65, 101, *102*
Vicars, John (1580–1652) 50
Vichy-Chamrond, Marie de (1697–1780) 80
virtuosi 52–3
Virtuoso, The (play) 53
Voigtländer 128
voyeurism *115*, 133

Walpole, Horace (1717–1797) 64, 80
War of the Worlds, The (book) 126
Watkins, Francis (1723–1791) 77, *78*
Waves, The (book) 144
Wells, H.G. (1866–1946) 122, 126
White, William Hale (1831–1913) 122
Wiesel, Johann (1583–1662) 35, *36*
Wilkins, John (1614–1672) 49
Willdey, George (*c.*1681–1739) 73
Woolf, Virginia (1882–1941) 144–5
world systems
– Aristotelian 16, 18, 30
– Copernican 18–19

– Ptolemaic 16–17
– Tychonic 19
Wren, Sir Christopher (1632–1723) 40, 57

X-rays 153, 163

Yeats, W.B. (1865–1939) 121
Yerkes, Charles (1837–1905) 142

young gentleman and lady's philosophy, The (book) 117

Zadkiel, *see* Morrison, R.N., Lieutenant R.J.
Zeiss 129
Zucchi, Niccolò (1586–1670) 56